WACONOMICS:

Redefining Water's Role in Our Economy

by Michael Zitron

© Copyright 2018 Michael Zitron

ISBN 978-1-63393-428-3

Author and Founder Michael D. Zitron
waconomics.org | wmmission.org

Author Photo by Nicholas J. Petriccione | www.bynicholasj.com

Comic Illustrations by Samuel H. Clayman | www.samclaymanpaint.com

Published by

210 60th Street
Virginia Beach, VA 23451
800-435-4811
www.koehlerbooks.com

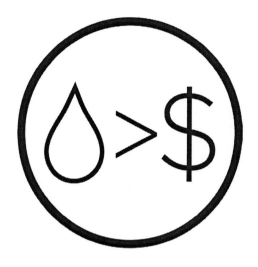

WACONOMICS

REDEFINING WATER'S
ROLE IN OUR ECONOMY

michael zitron

VIRGINIA BEACH
CAPE CHARLES

DISCLAIMER

No part of this book may be reproduced or transmitted in any form or by any means, electronic or mechanical, including photocopying, recording, or by any information storage and retrieval system, without written permission from the author.

The information provided within this book is for general informational purposes only. There are no representations or warranties, express or implied, about the completeness, accuracy, reliability, suitability, or availability with respect to the information or related graphics contained in this book for any purpose. Any use of this information is at your own risk. Discussions of specific products, companies, or technologies is for information only and is not intended to be an endorsement.

The information and conclusions within this book are derived from the author's personal thoughts and opinions. This book is not intended to be a complete and definitive overview of the field. It is intended to spur thought and discussion about various approaches and options for the use of water resources.

CONTENTS

Promise 1

WACONOMICS 101 3

CHAPTER 1: American Dream,
Meet Water. 18

CHAPTER 2: Marching for Water 32

CHAPTER 3: Water Stamps,
Before Food Stamps 41

CHAPTER 4: Made in the USA. 57

CHAPTER 5: Water Went Missing
from Money ♦. 72

CHAPTER 6: Apples to Apples ♦ 89

CHAPTER 7: Water is Volatility ♦ 98

CHAPTER 8: Global Conquest,
Corporate Commies ♦ 108

CHAPTER 9: History Repeats Itself ♦. 126

CHAPTER 10: Water Bubbles 139

CHAPTER 11: House of Cards
Floating on a Raft 149

Chapter 12: Why Fix the Pipes? 167

Chapter 13: I Went Undercover 176

CHAPTER 14: Wall Street's Pipes. 194

CHAPTER 15: The Candle of Profits
Burns at Both Ends. 208

CHAPTER 16: Win-Win-Win 228

CHAPTER 17: Quality Water for Every Tribe 247

Citations and Sources 253

PROMISE

Every dollar that *Wa-conomics* returns in profit directly goes to Water M. Mission, a nonprofit that stands up for water—that believes in World Water Day, every day! One hundred percent is allocated into a philanthropic water mutual fund to be used for effective projects and innovative water solutions around the globe. WM Mission was created to educate the fortunate, provide quality water for every nation and tribe, and use water as a catalyst for worldwide economic liberation.

When we buy this book, we give water to those without, and we improve the quality of our lives. When we read, we are closing the gap on gender inequality globally. As we learn and practice these conversations, we invest in our children, grandchildren, and beyond. And as we believe in price, we separate ourselves from all the limitations before us.

Water is constantly taking one for the team, absorbing every human pollution, with all its skin in the game but no voice to be heard. Because water is the ultimate martyr for all of us, I treat it as a public duty and responsibility to connect water's significance for everyone and to lead people toward water solutions, netting societal expansion, progression, and intrinsic value for citizens and corporations alike.

Please consider *Wa-conomics* science nonfiction. Feel free to jump around to what you are inspired to read. Chapters with water drops invite you on a journey back in time, onto foreign land, through war, and fast forward back to US soil and its economy. If the material seems mind-bending, then choose your own adventure, skip to the next section, or slide right to the next chapter. This is supposed to be fun, conversational, and never absolute. We must learn how water affects every corner of our lives to save our families' legacies and grow our money. Please step outside the box with me. Together we will come to understand water, one of the most powerful tools the future will ever need.

WAC 101

WATER IS IN EVERYTHING vital to our lives. Everything in our wallets makes more sense when we explain how water shapes our cash, credit, debts, healthcare plans, home values, and families. We use it to make food, livestock, the clothes we wear, and the microchips running every audio and visual form of media. Our technology is prepared first with water. Water is used to extract and process petroleum from the earth, and then refine that oil to gasoline for the cars that still run on fossil fuels, all of which are built from water-intensive metals and plastics. Water cools the reactors that generate nuclear power. We use it to heat our homes, to clean and cook, to shower, to flush, to wash our cars, to water our golf courses, to swim in pools, and to ski and snowboard on mountains. It's in our medicines, car tires, and in every basic building block of our homes and classrooms.

Water is the most upstream input to all life and productivity. It is the single most important substance to humankind—the staple of our food, clothing, and energy supply. The human brain is 90 percent water. Our bodies are 70 percent water and the surface of the earth the same. Water is what makes this blue planet so unique. Evolutionary scientists believe that we evolved from water. When we explore new planets, the first thing we look for is the possibility that a planet holds water. So why not look at the unaffordable mysteries behind the present-day cost of living for citizens, costs for corporations, and struggle by governments with water as our guide?

Water consumption is so vast and integral to our very existence that most people don't comprehend the depth of our dependence on it. Nor do they value it properly. Water rights now trade like gold on commodity exchanges. Investors and speculators prospect for water the way energy companies explore for oil and natural gas. The wealth and stability of nations rise and fall with the availability and cost of water. There is even evidence to suggest that fertility rates correlate to *water value.*

One pound of steak takes ten tons of water to create. We use 592 cups of water when we drink one cup of coffee. We need four to ten barrels of quality water to make one barrel of oil—which is used to power energy-intensive desalination plants. It takes 157 gallons of water to make a bag of sugar found in restaurants and pantries. According to Water.org, "an American taking a five-minute shower uses more water than the average person in a developing country slum uses for an entire day." The US Geological Survey estimates that each American uses up to 100 gallons of water per day, and some water scientists put the number at nearly twice that. The barrage of water consumption statistics is daunting, though some of the exact measurements for usage are certainly debatable.

What's shocking is not the human rate of water consumption, but how grossly we undervalue this mainstay of our very existence. Water's role is underappreciated. We treat it like air—freely breathed and inexhaustible. It's an afterthought. It's so low-cost in the United States that most consumers run the tap or flush without understanding the sacrifice water makes for us at its current price. In my hometown of Virginia Beach, we currently pay about $4.41 per thousand gallons. In many places in the US, water costs half that—less than one penny per gallon. Might we agree that water is taking one for the team and that the real value of water is disconnected from the perceived value?

So what's wrong with cheap water? Everything else costs a lot of money, so why pay more for something that helps us make those expensive items?

This question is a lot more complex than it may seem. And the answer to it even more so. To really understand the implications of water value, someone needs to bore deeply into water usage, government's interpretation of that usage, economic policies in the United States, and the global trade of commodities. None of the key economic indicators that gauge our inflation, the cost of living, or prices we pay, adequately measure our water value—or even include water quality. Why is that important? How did this happen? What are the implications?

Wa-conomics attempts to provide answers free of fear and generate simple solutions. Because we don't truly consider water's value, our economy is a house of cards upheld by layers of accounting tricks. Because water is in *everything,* we experience what I call *water inflation,* leading to a *water gap* in our households. The only way to close that gap is to put water back in our wallets, and fully recognize and acknowledge water value.

The principles of wa-conomics are as easy as 1-2-3.

1. Water is in everything. When a plant doesn't get enough water, it dies or doesn't bear enough food. When a manufacturing plant doesn't get enough water, everything in that sector gets expensive—and in 2011, it happened just that way in Chile. The first half of 2008 was another perfect example of the world being net short water and having no way to express that but through the price of oil and food. That is what I mean by *unrecognized water value.*

2. With water in everything, we experience water inflation. How did that happen?

    When the EPA came on the scene in the 1970s, it gave

water a higher value, but never explained it that way to the American people. Rather, the EPA slapped domestic manufacturing on the wrist for polluting without clearly communicating why. So the United States *hasn't*, as a nation of people, accurately recognized or valued the water we do have.

3. Why do we experience water inflation? Reason one: as our population grows, we have less *finite* freshwater per person (supply and demand). Reason two: water quality has diminished due to manufacturing (pollution, *negative externalities*, and *unrecognized leaks*). We can see that in the amount of life that water now supports. Since 1970, it's estimated that freshwater biodiversity has declined by 37 percent—with tropical water regions diminishing by 70 percent. Reason three: government benchmarks which measure inflation, like the Consumer Price Index (CPI) or Personal Consumption Expenditures (PCE), have less water in the data year after year, and additionally, there's no water quality index in those benchmarks. If the quality of our water declines and nothing links that reality with healthcare expenses, then we the people are the only ones on the hook. Finally, reason four: we lose and leak—like loose change—vast amounts of water through our pipes and we dismiss that lost water during the accounting process, as *non-revenue water*. Thus, households absorb water inflation over time, experiencing a water gap as water goes missing from our wallets.

4. So, we need to put water back in our money and close this mystery gap in our expenses, incomes, and economy. When we do this, we are going to harness water's power to revive economic health. We are going to reconcile water value appropriately for agriculture, industry, and residents (AIR).

In the end, we will have two choices about how the price of water will rise. Would we prefer to pay a toll to our country and invest in our family, or pay a toll to a multinational corporation? Will our decisions be on our terms or someone else's? The fate of the American dream depends on that choice.

$$\Diamond > \$$$

Since 2006, I have traveled the caverns of my mind to develop my views on the economics of water. The deeper I drilled into this topic, the more convincing the evidence became that water value is the cornerstone of economic prosperity and, conversely, poverty. In 2017, people marched for gender equality, among many other issues. It's worth noting that women's inequality is the very highest in parts of the world where water is not readily available, where access to water is absolutely limiting to life.

It's time to reveal what I have discovered, to raise awareness about the most precious resource we have and why undervaluing it could destabilize our economy and undermine our lifestyles and our families' ability to thrive. How I arrived at the theory of wa-conomics is unconventional and undoubtedly forward-looking. I make bold assertions and prescriptions that few, if any, elected leaders would dare broach. Boiled down to its essence, I argue in these pages that we need to value water adequately and then raise, rather than hike, water prices (water rates) and use our public water sector as a catalyst for long-term economic health in the US.

I have written thousands of pages on water economics and water's disconnect with prices. I condensed much of that research to provide a simple explanation of how water connects us all and how best to value water so we can enrich the American dream, our sovereignty, and lifestyles. We can give all employees a raise in real wages by closing the water gap. When we do this, we can build the middle class and, in doing so, elevate socioeconomic classes into new income brackets—without taking away from more advantaged groups. It is scarcity thinking to say that we must take away from one group to benefit another. *We can all rise with water value.*

All the data I studied and crunched since 2006 point to a chilling conclusion: hidden prices we pay for water are fueling incremental financial terror. We, as a society, incentivize consumption. Yet the price we pay for almost everything is, in both simple and complex ways, affected, manipulated, and subsidized by government agencies. The problem is that the reality of water value is out of whack with its force on the market.

Some parts of the world are running out of potable water, including swaths of our very own American Southwest. Yet corporations and investors are gobbling up water rights for future profits, and growers and ranchers are using vast amounts of North American public water to raise cattle and feed, then selling those water-rich commodities cheaply overseas. We are, essentially, exporting our most precious resource for pennies. We, as a consuming nation since the 1970s, have been importing aqua-rich products at a discounted rate from poor countries, depressing their economies, environments, and lifestyles. Whether it makes "cents" or not.

The smart money in this country sees the threat of water inflation and is positioned to profit from it. Unbeknownst to me, just when I began writing this book, water ETFs—like mutual funds—were created for the first time in the stock market, later justifying the timing of the cause. Farmers and ranchers get rights to cheap water because of government subsidies and farm support. Instead of using it, they sell it at a higher price to cities, companies, or investors—knowing the timing of water's importance, they can generate more money from those rights than its intended agricultural use. Corporations are bottling tens of millions of gallons of *public water* out of states like California—for a permit that costs less than $550—and selling bottled water for exponential profit. Big-money investors indirectly prospect around the globe for freshwater supplies by buying the land that sits over them. For the sake of this conversation, we could include well-known media owners, legendary investors, and presidential oil tycoons.

But the purpose of *Wa-conomics* isn't to demonize presidents, point fingers at politicians, end the need for water-related corporations, or harass ranchers and farmers who swallow vast water supplies or own water rights. That's in the past. We are moving forward. We create our future. Water inflation is not a Republican or Democratic issue. I'm not here to beat up on those who liberally flush and wash dishes without giving water usage a thought. I love water and love to use it. I'm not advocating that we stop eating meat, or eating almonds, or washing our cars, or taking showers. My arguments for change are economic—not emotional or political. It's about *price*, folks. That's one thing we can all agree on, because it's neutral. Aside from our loved ones, price is something we focus an infinite amount of our attention on: the price or balance of our online bank statements, the price of our investment accounts, the price of gas, the price of our power bills, the price of groceries, and work, the sacrifice—or price—of our time.

As you've probably figured out by now, this book is titled *Wa-conomics* as shorthand for simple water economics. But it could easily be called *Whack-onomics,* because everything about our economy is whacky when we don't fully recognize water value.

What do I mean? Consider that we waste roughly 30 percent of the water we treat and pump into businesses and houses through antiquated and broken infrastructure. Big banks, government engineers, and NGOs have estimated that between *10 and 50 percent* of treated water is *lost* through pipes in the United States alone. That means for every hundred gallons flowing underground, about thirty leak through cracks, holes, loose connections, metering miscalculations, and theft. *It's called non-revenue water (NRW), and we don't account for it on our water bills.* The US government prefers to declare *average* daily losses at 16 percent, but as we'll realize through *Waconomics,* removing the noise of highs and lows doesn't benefit the people. Non-revenue water is the giant elephant in the room; there's no certainty behind government statistics, and for years NRW was not measured by many US states. *If our bank accounts and our personal assets—all made of water—leaked 10 to 50 percent of their value every day, and no one knew which account leaked what, would this be something we paid attention to and called the banks out on?*

The US brand, the dollar, is determined by our country's efficiency and productivity. Water is the most upstream input to all productivity. If we don't *account* for the tremendous leakage from non-revenue water, then our dollar is literally leaking from our paychecks—about 30 cents on the dollar. That's water inflation! And it's the reason our income cannot seem to keep pace with our expenses. Losing this much water is economic suicide. It is creating scapegoats for change, slowly shifting control of our water resources to private-based solutions instead of keeping it a *public duty*. If we don't act, those who control water will. We will be at their mercy, paying whatever water ownership demands.

Leaky pipes are no secret. The EPA has estimated that it would cost almost $1 trillion to fix and upgrade waterlines throughout the country over the next twenty years. And it's not just the underground pipes from treatment plants losing water. Households leak a trillion gallons annually—enough water to supply eleven million homes a year, according to the EPA. Do we often discuss swimming pools that leak and overflow? These are not sexy subjects, but if we had more of a connection with our money and water, they would be.

Numbers such as these are so amorphous that they're almost meaningless and we tend to block them out. What is real, though, are the strained budgets of local water municipalities struggling to get water systems operating efficiently, especially in older industrial towns that laid iron pipe back in the 1930s and 1940s. According to the *New York Times*, Syracuse, New York had four hundred water main breaks in a year, and New York needs to spend almost $40 billion to fix pipes around the state.

Few people dispute the need for a massive water infrastructure upgrade. Presidents run on the promise of a massive infrastructure improvement effort—but for some reason that never really happens once they're elected. Our bridges, roads, and pipes for water and wastewater are crumbling, especially in our older cities and communities. This is a common theme in state and local elections. Not only are tired pipes and outdated water treatment plants an economic issue, but they are personally taxing to our wealth and health. Water makes us sick and is the number one reason for death. Pandemics dwindled, populations surged, and life expectancy rose because of water pipes and proper infrastructure for wastewater. But when overlooked, water value can lead to serious injustices. The film *Erin Brockovich* and the person who inspired it are famous for shining a beacon on this issue.

The fiasco in Flint, Michigan, was a stark reminder of what can go wrong. Unfortunately, there are and will be many Flint, Michigans—more than we're ever likely to discover.

Why couldn't Flint pass along a higher cost for a better outcome? The real problem is an aged, broken water supply system in desperate need of an upgrade. Why isn't the money there to pay for such an upgrade?

The political problem is that raising water rates or taxes to pay for massive water infrastructure repairs is unpopular—or so "they" say. The culture tells us that no one is likely to get elected to office or appointed to a water commission on a slogan of raising taxes or monthly water bills. I felt that tension firsthand when working at the Hampton Roads Sanitation District in coastal Virginia.

At some point, the leaky pipes will be too old to patch, and water itself will be more sensationalized and appear scarce to the masses. When those two dynamics converge, we're going to have a perfect storm of rapid and steep water inflation. Politicians, who have been told to keep water rates low for decades, are standing on a powder keg, like Flint. Political acceptance of the city-run municipality will decline, and privatization will be a knee-jerk reaction.

Will politicians sell out state-owned assets and contracts to corporations under these conditions? *Will* the local municipalities be considered failing then and privatization "the solution?" At such a time, instead of slowly increasing the water rate under the local government's control, *will* we allow private corporations—with no allegiance to the citizens—to dramatically *hike* rates for the same water?

We could have had slowly rising water rates, on our terms. But no one is giving us the choice between the two. If privatization happens, we will have to pony up the dollars for attorneys in court to fight the private water companies. Over time, we could face startling price increases on monthly household water bills, and at the supermarket, gas stations, and online stores. Remember, water is in everything, and price hikes will run downstream to create massive water inflation everywhere. However, if we increase water rates on our terms, we can empower ourselves, our corporations, and our country.

My purpose here isn't to scare anyone. Rather, I want to raise awareness and make a case that we can avert any more financial regression by increasing water rates now. This message isn't one of doom, but rather of hope and confidence that we can have a sound and sustainable water supply and delivery system if we allow economics—instead of politics—to do its job. Accurately valuing water will stabilize our water supply system and reduce water inflation. It will, over time, translate to better job markets and stock markets, which improve the quality of life for citizens and corporations alike. Repairing our water systems now will be like giving ourselves, our corporations, and our country a raise later.

Exploring and explaining the dynamics of water economics is a complex endeavor. It goes way beyond supply and demand and simple ideas like water scarcity. This book is not about that. We can make more potable water through desalination, but at what price? And who controls that water? The trick will be properly valuing that water, which is exactly the challenge we now face. And that task could rest with special interest groups who would assign a profitable value for private corporations—we don't want that.

I will explore how water's true value is not properly reflected in key economic benchmarks that affect the dollars in our wallets—like the Consumer Price Index (CPI). I will show how water is responsible for the nation's gross domestic product (GDP). I'll explain as clearly as I can why water is, effectively, our truest *current-cy* or force—not a currency. Yes, my theories are outside the box, but when brought back into conventional economics, they are fiscally sound. Admittedly, I am not a university professor or a government economist. And it's important that I'm not, so that we can remain objective and have fun.

Confronting water economics, in its rawest form, is inevitable. My question is this: do we spend some incremental pennies now and confront the problem, or do we keep plugging and fixing leaks and selling water for less than a penny a gallon until the system fails our families?

$$\Diamond > \$$$

Since 2006, I have studied water economics as a father, homeowner, international business student, employee of a water municipality, market analyst, chemical commodity trader, and day trader. Each day, I sat in front of five blinking computer screens while I tweaked theory, wrote, and edited. I tracked movements in commodities and climate, and used indicators connected to funny-looking candlesticks and stock charts that ripped and roared like waves on the Atlantic. I compiled research from around the world on this topic, studying and analyzing trends in water bills, weather, and technological improvements in the sector. I have enough research and writings to fill three books the length of *Waconomics*. But for this effort, I decided to keep it simple and, I hope, understandable.

I promise not to riddle you with obscure studies, stacks of footnotes, and dense calculations. But this broad topic does force me to delve into tricky realities of fiscal and monetary policy and, more broadly, global trade and history. We can't fully discuss water value without providing facts and, yes, numbers, and where those numbers came from.

Most people who met me during the process of writing wondered, *Who's this guy, and what's his obsession with water?*

Always a poet but not always with purpose, I became a global citizen while studying at three universities: the University of Tennessee at Knoxville, Virginia Polytechnic Institute, and Old Dominion University. The beginnings of this book hatched while I was still a student exploring various political and economic theories. Over time, I realized that water was the catalyst for global wealth and poverty, growth, or repression. It occurred to me that we faced an international imperative—to innovate our water systems to raise our standards of living, and create access to freshwater for developing nations. I believe it is more crucial than ever to pipe international economic equilibrium, with water value as the cornerstone.

This moral imperative, and my passion for studying water usage and supply, flow from my background and upbringing. Motivation was a fabric in the tapestry of my family. My parents dug deep for success— they were spitting images of Type A, middle-class Americans. They built a dream home on the Lynnhaven River in Virginia Beach, then moved to the city's edge on the Atlantic Ocean, and eventually onto the Chesapeake Bay. They often shared stories of where I came from and what was expected of me because of it. I first complied, and later questioned who I was, but could not avoid the truth of my family history.

Growing up, I often heard the story of an only child—a tall young man with bright blue eyes. After getting rejected by the V-12 Program for being Jewish, that boy left high school to fight for his country in World War II regardless. He endured countless typhoons in the Pacific aboard the USS *South Dakota*, a.k.a. Battleship "X." He survived dozens of sunken ships and thousands of capsized lives and mutilated bodies. Eventually, before the war ended, the V-12 rightfully invited this hero into their Navy College Training Program. Upon his return from battle, he attended Princeton, Trinity College, and then Yale to study optometry and take over the family practice, where he reconnected with his high school sweetheart and wed her. He became a bar owner, studying during the day, pushing cocktails at night with his brother-in-law, and then returning home to his growing family and widowed mother. Dr. Richard and Elayne Silvers, my grandparents, found the American dream and set the tone for what I would become.

The seeds of my eventual calling took root at the University of Tennessee, with Alexis de Tocqueville. Later, while I attended Virginia Tech in Blacksburg, my intellectual fire was stoked by the political theories in *The Coming Anarchy*. Two ten-page research assignments converged to fifty pages, bridging commonality between Native Americans, Africans, and Jews. I attempted to connect different periods of history that had the same outcome, and build global citizenry through my words. In a sense, that's what I'll be doing here—connecting a global economy through water, our most powerful resource.

I no longer wrote for the school, and began a natural journey as a salmon swimming against the current. I was pooling streams of reading material. Running away with recommended textbooks, I was articulating ideas against the grain of academia.

I paused as a degree-seeking student, yet continued to embrace education. Just like my grandfather stopped and started his career path, I went home to defend my new commitment. I officially transferred to Old Dominion University, and would go on to graduate as Outstanding International Business Student with *Summa Cum Laude* regards.

I had become fascinated with water-related events in Chile. Back in the early 1970s, following a curiously funded coup d'état and economic intervention in Chile, the Chileans privatized their water systems. Other countries, including the US, had been meddling in Chile because of what I saw as a water paradox, not just a Cold War conversation or solely a dispute over natural resources like copper. I came to believe that wars were fought over the precious liquid more frequently after World War II than most people acknowledged. Water had more to do with global relations than the gold standard. Standards of living were tied to water—not precious metals. Floating currencies only further masked the real value of water. Nations prospered or collapsed because of freshwater supplies. Those who controlled water controlled wealth. Those lessons were clear.

While at ODU, I propelled my career toward waconomics. I went to work in areas that had everything to do with water. I essentially created a position at the Hampton Roads Sanitation District in 2010, becoming a specific project analyst in the accounting and finance division. The CFO thought it was cute when he called me "the intern." I studied the facts around non-revenue water and the implications of an antiquated, leaky water infrastructure. I went to HRSD because I wanted to substantiate a few theories. If I couldn't, this book would never go to print.

At that time, the burden to continue felt overwhelming. I was hoping to be proven wrong, not because of my sacrifice, but for the sake of the American public. After I received the stories and truths I came for, my "internship" was over. But HRSD was not enough for me. I had to get closer to the corporations tied to water. During this unrelenting journey toward waconomics, I wanted to get inside the boardrooms. So I served as an international chemical commodity trader for Quad and Tamaya Chemical, with operations in the US, Chile, and Belgium. Through graduation, I worked as a trader for Quad, a research analyst for Tamaya, and served at Business Development International. As such, I have watched the ebbs and flows as water-intensive products trade and react to water.

These positions were never for money, nor did they exist before my initial conversations or interviews. I was going after stories, or openly attacking wa-conomic theories, and turning down amazing opportunities along the way.

In that period, droughts, publications, news stories, and water-related events changed industries and redefined those markets. I decoded and timed droughts and their relationship to commodity prices, currencies, equity markets, goods and services, raw materials, and interest rate patterns. These economic concepts became simple yet bold characters in the water story. I back-tested and supported the hypothesis through documented experiences, news, and the price of commodities and stocks of the corporations who procured those resources—over and over until it was clear that no other explanation fit as consistently as water.

I learned to trade the markets, and studied Japanese candlesticks and chart patterns for this cause. I tracked commodity breakouts during droughts, and then compared monthly statements by the Federal Reserve about inflation. The best validations later came from investment analysts covering corporate models and their strategy toward profiting on water. I followed corporations: their stock prices, their costs, and their inputs. When we follow the money and where it's moving, we find prices and, more importantly, costs and motives.

The value of water got to me more as a husband, father, provider for my family, and global citizen. Aiming to be a middle-class family man and taxpayer, I pondered our inevitable connection to water and our economy. It affected all my decisions as a parent—which neighborhoods to live in, schools for children, access to quality foods, and convenience. It occurred to me that my research had real-life implications. Water was in everything. Not just now, but in the past, and in the future. Water is Main Street's issues, not just Wall Street's secrets.

What types of human-made chemicals were in the water that grew an apple and what are the associated health risks? What would it cost without all the chemicals and least-cost rules applied? Does our government calculate inflation based on the cost of the apples we choose to buy from the grocery store or local farmers market? How much is my water bill? How much should it be? Is milk accurately priced? If not, how much is it subsidized? What should we pay for a gallon of gas? What happens to our purchasing power if drought pummels California and Texas? What happens when more than one public water municipality goes bankrupt at the same time, and more cities like Birmingham, Alabama, and Flint, Michigan get the attention of the news? Are we prepared to understand that?

Please step outside the box with me, and understand why water is one of the most powerful tools the future will ever need.

CHAPTER 1

American Dream, Meet Water

SO, JUST HOW DEPENDENT are we on water? Consider this: without adequate water supplies, many of us couldn't switch on the lights in our homes.

While ignorance is bliss, not everyone sees the reason to care about water yet. Take away the water from the kitchen sink because we're told, "It's no longer good," and they might. Kill the water supply for energy to heat and cool our bedroom, from the bathroom toilet, and where we sleep no longer feels like home. If we need a financial advisor to plan for retirement, then we need to be advocates for water now.

"It takes 37 gallons of water to grow, package, and ship enough coffee to make a single cup of morning Joe," wrote Nick Hodge in *Wealth Daily*. That equates to 59,400 percent more water than coffee. One ton of wheat requires a thousand tons of water. That's 100,000 percent more water than wheat. It takes two thousand gallons of water to make one gallon of milk. Obviously, cows aren't taking long showers. They're simply eating water-intensive feed and dyeing the surface waters crimson when slaughtered with cleanup costs.

Regarding coffee cups and wheat tons, where our mystery expense lies today is what we can call *unrecognized water value*. Because more world populations are buying bread and coffee, water is why everything is more expensive.

The water requirements of personal consumption convey the message that global trade is all about exporting water, a.k.a. aqua-exportation:

- 99 liters of water per apple
- 107 liters of water per banana
- 10 liters of water per rose in water-scarce Kenya

When we import these items, they come with "strings attached." Some of those strings could be environmental degradation to the country growing the apples, the toxic chemicals inherited in those bananas we consumed, and wars in Africa for roses.

What about when we export water? Look at the Colorado River, the so-called "American Nile." It maintains the life of thirty million citizens with its tributaries. The Colorado offers drinking water, the power from its flow helps create electricity throughout seven states, and leftover runoff aids Mexico's water shortages. Western ranchers support some five million head on 250 million acres of US land. According to Wade Davis, these cattle, 10 percent of the national herd, require nearly half the flow of the Colorado, more than ten times the water used by all the area's cities and industries. Much of this meat, using our limited water supply, is exported outside the United States. Selling meat overseas is profitable for cattle ranchers, but what about everyone else dependent on the Colorado? What happens to the price of houses and other real estate if this river ends up depleted?

We sacrifice our land value when we sell that water for the price of steaks to Saudi Arabia. We send water-consuming *aqua-exports* to the desert; therefore, global populations push demand for goods and services where water is scarce and where steaks are typically not. It's like pushing a rope. While US dairy reaches every corner of the globe, governments around the world subsidize the dairy industry by as much as 600 percent—entitlement or welfare for meat and milk in Saudi Arabia. If the real cost of water were factored into the price of cattle or milk, other countries would be paying exorbitantly for American beef and dairy. Either that or they wouldn't buy at all. The point is that water is inaccurately priced almost everywhere in the world. Its current price structure is false.

Even innovative environmental programs and water efficiency plans by corporations are aqua-exports. How is that? Integrating new, less water-intensive processes and using water-less production techniques represent aqua-export costs in the form of capital investments, time, and company energy. Money is an aqua-export because it is exchanged for other aqua-exports. Everything is essentially an aqua-export, because water is in everything.

All countries import and export goods and services. Because commodities, food, and raw materials use tremendous water, they are considered now as aqua-exports. When produced, these consumer products are embedded with and require hundreds-to-thousands of percent fresh, potable, and treated water. Through water-intensive product development, aqua-exports become value-added goods and services, which *should* retain what we can call *water value*. Aqua-exports also emit controversial and costly discharge or pollution in their process of upstream-to-downstream manufacturing. Total water value must account for the cost of cleaning up contaminated and compromised water. Aqua-exports *should* cumulatively represent water consumption, the penalty pollution (*negative externalities*), and the leftover social costs (*unrecognized leaks*) because, in the end, it all changes the real price of goods and services for people. Aqua-exports don't, which in turn discounts our purchasing power and financial freedom later. Why are we losing out? It's below the surface level. It's in the water value.

When we think of water value in aqua-exports, negative externalities (*NEs*) are our most tangible voice because there is costly evidence everywhere we expand, grow, output, and produce. NEs scream taxation and unrecognized leaks by irresponsible or unbalanced corporate advent. NEs are a cost or tax upon us—the cost of relentless wants (marketing) and growth (consumption) that will ultimately lead to higher prices, regulation, and taxes. As we consume, we demand aqua-exports, so effluents are then created and later cleaned up in response (new costs and taxes)—further increasing the total price structure and value of goods. Guess who pays for it all?

Economists describe NEs when a corporation understates the total cost of their production by polluting a lake, not entirely cleaning up their presence, and thus transferring a social cost to the community, a.k.a. unrecognized leaks. Why are *ULs* a social cost? Because we pay for them, without anyone else accounting for that cost of living on our behalf—a stealth or hidden inflation on us. The corporation benefits from lower production costs since the firm was not entirely accountable for their actions. Eventually, states or the Environmental Protection Agency (EPA), charges the corporation for a portion of their NEs—their effluent discharge (cleanup costs, carbon credits, etc.). We as taxpayers fund those government efforts and programs through taxation—a fee found deep within our paycheck. The corporation later rolls the additional cost of cleanup or environmental improvements in operations into *our prices*. Any and all costs increase the price we pay. The remaining pollution (non-penalized effluents) are ULs—some reduction in the value of our property, health, and quality of life. The sum of this cost is a stealth tax, revved up *water inflation*, but overlooked in most aqua-export values.

When NEs are addressed appropriately, their costs can lead to positive externalities (*PEs*), offsetting social costs or creating greater social benefits. NEs are like a cost or tax on both corporations and citizens. Moreover, corporate costs and taxes can lead to PEs, such as clean bodies of water to drink and swim in (EPA), safe and protected cities (US military expenditures), sound road systems (gasoline tax), and dedicated teachings through public schools (property tax). This is the spillover benefit from managing the public's well-being. In evaluating our water value, there are both negative and positive externalities. NEs are simply a vehicle we will harness to justify higher transparent prices for water. PEs are the result of investing time and money (tax dollars) into water operations so that they become efficient, become optimal for the body, and strengthen our purchasing power. Within the United States, any tax either addresses or presents a negative or positive externality to all citizens—even free-riders.

Unfortunately, not all NEs are captured, and not all parties involved with water's use worldwide are regulated equally, which, in turn, changes the game of life. There is a social cost of ULs that lasts for generations. Whilst US environmental regulation is far from perfect, we eat or drink aqua-exports from other countries with contaminated groundwater and inferior water quality. Those nations' regulation is minimal or nonexistent. On the surface, everything seems fine, but the ULs of water quality are like a ticking time bomb in our bodies, striking children and families unexpectedly. Most US citizens don't connect with "Agent Orange" used in the Vietnam War in the 1970s. But the Aspen Institute notes that the level of dioxin connected to Agent Orange "is a highly toxic and persistent organic pollutant linked to cancers, diabetes, birth defects and other disabilities." While the people of Vietnam directly suffer, as seen below, the coffee we drink in the US grows in Vietnamese regions, where dioxin remains a part of the ecosystem. "Dioxin buried or leached under the surface or deep in the sediments of rivers or other bodies of water can have a half-life of more than 100 years." Vietnam's coffee is an aqua-export shipped to the US and sold in our cafes with strings attached—the ULs are a *potential* health risk, a water inflation on our lives.

Le Van Dan (C) looks at his disabled grandson Le Van Tam (L) as his daughter feeds another sick grandson in their family house in Phuoc Thai village, outside Danang April 12, 2015. Le Van Dan, a former artillery soldier with the South Vietnamese army, said he was exposed to Agent Orange more than once, including being directly sprayed by U.S. planes near his village before he joined the military. Health officials confirmed two of his grandsons' disabilities are due to his exposure to the defoliant, Le Van Dan said. REUTERS/Damir Sagolj

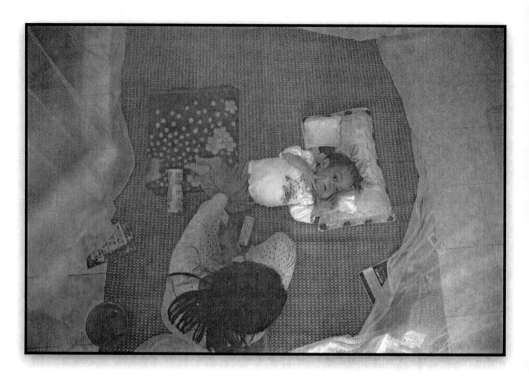

7-year-old Nguyen Van Tuan Tu, who suffers from serious health problems, is looked after by a family member under a mosquito net in their house near the airport in Danang, in central Vietnam April 12, 2015. When Nguyen Van Tuan Tu's father started working at Danang International Airport in 1997, he was not aware of the health risks associated with Agent Orange and he collected fish and snails from a contaminated lake nearby for the family to eat. His first child to be born after he started working at the airport, was born in 2000 and died in 2007. Nguyen Van Tuan Tu was born in 2008 with same symptoms as his late sister and doctors and parents believe their health problems are linked to effects of Agent Orange. The couple have one healthy daughter who was born in 1995, before they started working at the airport, and she is now a university student. Danang airport was a U.S. airbase during Vietnam war and since 2012 both the U.S. and Vietnam are conducting a clean-up operation at the site. REUTERS/Damir Sagolj

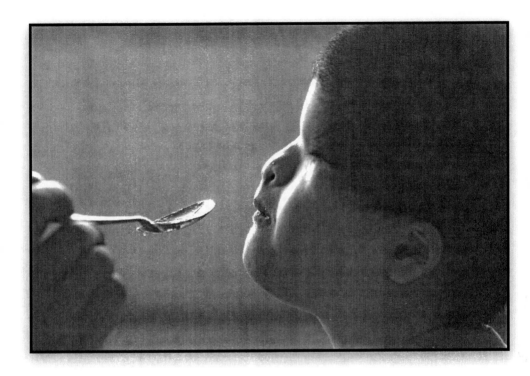

12-year-old Tran Huynh Thuong Sinh is fed by a hospital staff member at the Peace Village in Tu Du hospital in Ho Chi Minh City April 14, 2015. Tran Huynh Thuong Sinh, whose parents and grandfather were all exposed to Agent Orange, was born without eyes and with other serious health conditions. Officials at the hospital link his health problems with exposure to the defoliant. According to the head of the Peace Village, more than two-thirds of its over 60 patients are from areas that were heavily sprayed by Agent Orange and their health conditions are linked to the use of the defoliant. REUTERS/ Damir Sagolj

◇>$

As the international economy expands, the push from populations demands more machines to build infrastructure. By some older estimates, it takes about forty thousand gallons of freshwater to manufacture a car and sixty thousand gallons for a ton of steel. Imagine the volumes of water required in the fabrication process of agricultural and mining vehicles. Imagine how much diesel, gas, and refined petroleum they burn. Have you seen how big these machines are?

In 2009, the *average* US industry estimate for drilling and hydraulic fracturing required 7.6 barrels of water to yield one barrel of oil. We need roughly seven barrels of water for one barrel of crude? That is 760 percent more water than oil! In California, a barrel of oil requires 10.5 barrels of water. The oil and gas industry calls this a *WOR*, or a water-to-oil ratio. After we procure the crude, we then refine it to gasoline, where we need another round of water—roughly 250 percent more per gallon. Can we now imagine how filling up our gas tanks is an act of buying aqua-exports? The more we drive, the more we develop our dependence on water.

Aqua-exports take the shape of water-intensive services, such as our electricity, accompanied by unknown baggage and more strings attached. In this circumstance, we are discussing the water used to orchestrate oil and gas production, which results in *produced water*. As we now realize the nexus between energy production and water value, we find an associated private cost for taking water and an opportunity cost when polluting it. Today, power companies are now switching from coal-fired technology to natural gas alternatives. Natural gas combustion takes place in boilers and turbines to generate electricity. Yet this so-called "clean energy" to heat and cool the home leaves exponential hazardous waste behind in our aquifers. Two separate studies agreed that this was "the largest single waste stream in the US" (Allen and Rosselot in 1994 and Graham, Bowman, Katz, and Kinney in 2004; see Citations and Sources section).

The water may look the same today, but after produced water invades our home—on the surface alone—dishwashers send produced water vapors into the air, and we breathe toxic hydrocarbons such as benzene. According to Dr. Teitelbaum, certified physician and medical toxicologist from Colorado, our "dose of the volatile organic compound from the shower water will be several times the dose you would have gotten from the drinking water." And below the surface, there is a network of evidence suggesting our food and livestock are grown with this waste stream. Even if we do not feel, see, or taste that produced water as we pay for our electricity and groceries, water is in everything, and connects us all, one way or another.

Furthermore, all transported water and produced water from oil and gas production (*O&G*) becomes another form of baggage, with additional travel fees—compounding layers of cost with fossil fuels and the total meaning of aqua-exports. The electricity is a service, but requires tremendous water and creates pollutions in the process. Therefore, our power bill paid online represents digital aqua-exports. Can we now see how natural gas is an aqua-export, developing the meaning of water value?

One semiconductor needs three thousand gallons of water (think electricity). It takes thirty-two liters of water to make a single microchip (think computers). So, our digital lifestyle and social media are prepared first with water. Nick Hodge, an investment strategist, makes it crystal clear: "It doesn't matter what type of energy you're talking about ... [F] rom the deepest oil shale to the newest solar panel, it needs water."

Water supports *AIR* wholeheartedly: *agriculture* (which accounts for 70 percent of all water consumption), *industry* (20 percent), and *residents* (10 percent). Industry's consumption is mostly through the production of chemicals, drilling oil and gas, and mining metals. As the world's population grows, each sector of *AIR* is growing out of its shoes and leaving larger water footprints. According to the United Nations, the planet faces a 40 percent shortfall in water supplies by 2030 because of urbanization, population growth, and increasing demand for the water required to produce food, energy, and industry. We say, "Give it

140 percent," but can water really do that without drastically shrinking our buying power? Wouldn't everything get a lot more expensive under these conditions?

Water is "energy's energy." The three largest industries of the world, in order, are oil, electricity, and water. If the top two sectors do not exist or perform without water resources, then it's a no-brainer: the creation of food and energy is explained by access to water and water's cost. Water as a resource trumps all, and it too is an aqua-export. Our basic needs need water.

Think about water's limited supply. Think about the demands we put on water with everything in our lives. The quality of our water shapes our lifestyles. Supply and demand should determine a fair price. Are we sure that the water we use to make everything we own is going for the right price?

The story of underpriced gasoline might help explain why we live in a house of cards floating on a raft. Shockingly, gasoline prices are too low even when they feel expensive. Author and Worldwatch founder Lester Brown declares there was a massive *market failure* in the US during 2007 when gasoline was $3 per gallon. But even at that level, the price reflected only the cost of discovering the oil, bringing it to the surface, refining it into gasoline, and delivering the gas to service stations. It fails to notice the costs of NEs discussed above, the tax subsidies to the oil industry (such as an oil depletion allowance), the military expense of protecting oil access in the politically unstable Middle East, and the healthcare costs for treating ULs.

So, what does a gallon of gas really cost? Brown cites a study from the International Center for Technology Assessment: "These costs now total nearly $12 per gallon ($3.17 per liter) of gasoline burned in the United States. If these were added to the $3 cost of gasoline itself, motorists would pay $15 a gallon for gas at the pump. In reality, burning gasoline is very costly, but the market tells us it is cheap, thus grossly distorting the structure of the economy."

Gas is undervalued because its main driver for extraction—water—

is not fully recognized or valued. That gross miscalculation persists today in everything we consume, including our time. Time is money, and our families depend on both.

○>$

Before we move forward with our families, let's do some simple math. At one time in history, there were one billion people on this planet with a fixed amount of freshwater to serve them—water was cleaner then and didn't require regulation to clean it (think zero costs). Fast forward to a time when there are seven or even ten billion people on this planet with the same, fixed amount of water, but this time much of the water that was once fresh is contaminated (think manufacturing and more people's waste). We have to pay to clean and regulate that water (think lots of time and money). According to the World Wildlife Fund, it's estimated that freshwater ecosystems have declined by 37 percent since 1970—with tropical water regions diminishing by 70 percent. That means we have a whole lot more people asking for relatively less and less freshwater. So why wouldn't everything get more expensive?

To make freshwater seem even more valuable, we constantly have to contaminate the freshwater we have left with chemicals to grow food, build houses, and create transportation and energy—all while we have to take care of the water when we are done making those items. If tremendous water was used and affected in the creation of each good and service, then it's so very important to imagine each good and service as an aqua-export, with lots of water value. If we can visualize the water value required to make goods and services, the penalty pollution to water, and the leftover unrecognized leaks affecting our bodies, then it increases the real price of goods and services for people. We must value goods and services and see them as aqua-exports, because water remembers. It remembers its freshwater status, its obligation to manufacturing, and then its contamination. Water is in everything, and connects us all, one way or another.

Adding Up the Value of Water
Breaking Down the Real Costs of Aqua-exports

Fossil Fuel Example

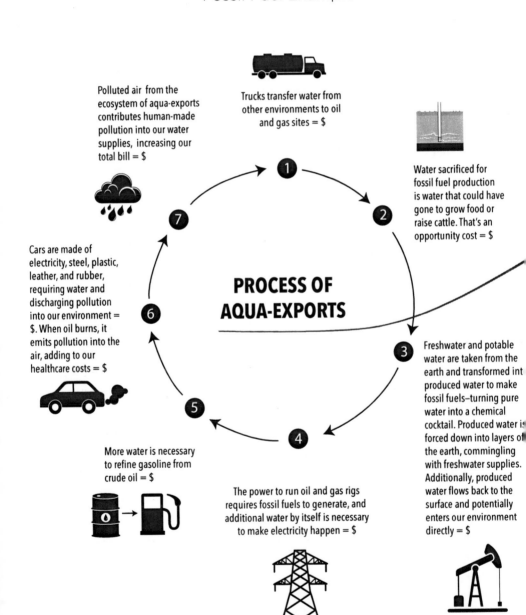

Trucks transfer water from other environments to oil and gas sites = $

Polluted air from the ecosystem of aqua-exports contributes human-made pollution into our water supplies, increasing our total bill = $

Water sacrificed for fossil fuel production is water that could have gone to grow food or raise cattle. That's an opportunity cost = $

Cars are made of electricity, steel, plastic, leather, and rubber, requiring water and discharging pollution into our environment = $. When oil burns, it emits pollution into the air, adding to our healthcare costs = $

PROCESS OF AQUA-EXPORTS

Freshwater and potable water are taken from the earth and transformed int produced water to make fossil fuels–turning pure water into a chemical cocktail. Produced water is forced down into layers of the earth, commingling with freshwater supplies. Additionally, produced water flows back to the surface and potentially enters our environment directly = $

More water is necessary to refine gasoline from crude oil = $

The power to run oil and gas rigs requires fossil fuels to generate, and additional water by itself is necessary to make electricity happen = $

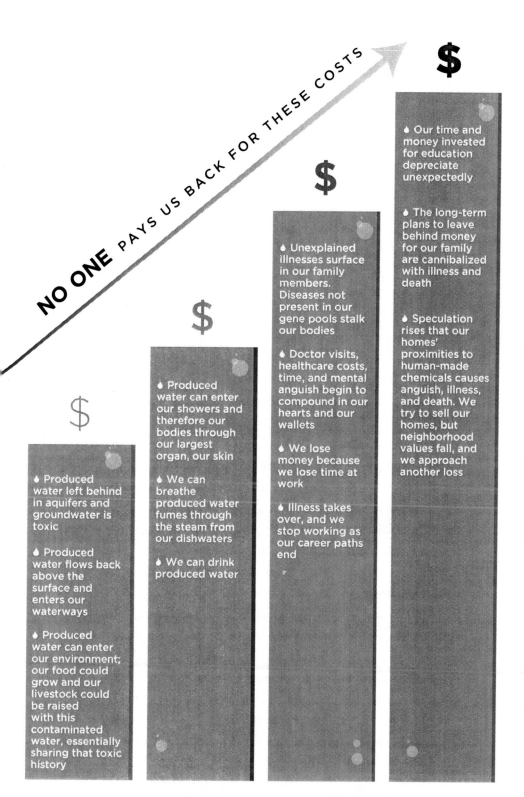

Examples of Water Inflation

CHAPTER 2

Marching for Water

WHAT IF I TOLD you that water was the reason kids were so expensive? Would you believe me?

The total fertility rate in the United States peaked at about 3.7 births per woman in the late 1950s. By 2010, it dropped to 1.9.

A couple bears enough children to replace themselves at 2.1; anything below that means the population is dying. Why are we not capable of replacing ourselves? Before the 1950s, the social expectation was that a man's salary could still support the entire *nuclear family*. But that is not the case anymore. Many married women and mothers now work outside the home.

After World War II, the average US family initially had more children because *life* offered the global population sufficient water inputs for consumer products. Thus, wages equaled needs. As populations exploded in developing nations and demanded water, US materialism met marketing in the 1950s, 1960s, and 1970s, and something was very different after the EPA.

Back in the day, infrastructure for clean water and sanitation were partially responsible for the population surging in the US during the

1800s. However, the importance of a woman's income on a household's finances was not. It began after the Federal Water Pollution Control Act of 1948 and accelerated following the onset of the 1970s. Before environmentalists recognized water value, a woman between the ages of 15 and 44 had about four kids, and one-income households were standard. Today, is the shrinking nuclear family a response to the quietly rising cost of water in everything they consume, even as both adults have joined the workforce?

When the EPA came onto the scene in the 1970s, it gave water a higher value, but never explained that to the American people. *We didn't get the memo.* At the time, the message the media perpetuated was that the EPA was slapping US manufacturing on the wrist for everything associated with pollution. It was as though the EPA was punishing Detroit, Pittsburgh, and the cities of West Virginia. The reality was that anything and everything that needed water was getting more expensive because the water had more value. Salaries and wages weakened against water value, US buying power shrank, aqua-exports got more expensive, time became much more valuable, and fertility responded. That's when the American dream started to become so expensive. Despite the rise of dual-income families, the twenty-first century is seeing a declining middle class and declining fertility rate. Boiled down, the US is poorer than it was half a century ago in purchasing power, family size, and family time. When the value of water rose after the 1970s, something had to give.

Conventional wisdom says that after World War II, women have continually moved toward independence by joining the workforce. Women's passions redirected dollars away from time at home toward education, a climb for success, and household income. They reshaped family focus from quantity of children to quality of children. Women chose to delay child-rearing, shortening the window of opportunity for raising children, and reducing the total fertility rate.

However, women's income was not nearly as relevant to the financial stability of families until water had a recognized value to US environmentalists. Prior to water value, many women opted to stay home, raise children, and join them after school because they could. When that was no longer feasible, many families traded number of children for hours at work.

Today, we rationalize families having fewer children because everyone is justifying their career. We traded off larger families for smaller ones, granting ourselves more choices, freedom, and money. Instead of simplifying everything down to inspiration and choices, why not also consider water's role in the economy and what happened as water value rose through regulation?

All we hear nowadays is that there's not enough time in the day. Members of the typical US household openly admit they have far too little time on their hands. *And we all know that time is money.* Couples were fully replacing themselves in the 1950s, with many women staying home, yet today both parents work, women are having half the number of children, and quality *family time* is shrinking in lockstep. If a US citizen is constantly complaining that there's not enough time in a day for themselves or family, is that not also an admission that there's not enough money for their needs? Despite all the work, most adults in developed economies feel that between total daily and lifetime expenses, we are *still* awash in concerns about providing educations for our children, healthcare expenses, and retirement goals.

So, where is the payoff from cannibalizing the size of families and quality family time? There is no payoff, because this shortage of money and time is water inflation eating away at the quality of our dreams. Not that many people have noticed water inflation. Most of us instead simply joke around about needing a few more hours in a day.

This is not an absolute. It's a choice with household conditions that vary. But given a choice, wouldn't it be healthier for families if one parent—Mom or Dad—could stay home and raise their offspring before primary education? Isn't buying a home and building a family the American dream?

As developed economies matured, productive capacities were strained by water resources, and population growth was confined in the natural scope of progress. During World War II, women went to work for their country; throughout the 1960s and 1970s, ambition led women to the workforce; in the twenty-first century, society and the economy say that women need to work—as though it's no longer a choice. Families are shrinking, everyone is working, and everything is more expensive.

$$\lozenge > \$$$

Europe's empire was invasive, but why? Primarily because Europeans had matured beyond their own resources first and sought luxury. They did not have sufficient climate, land, or water for their growing, wanting population. Demand outstripped Europe's capacity. The quality of their standard of living became a sacrifice for other nodes on the globe, where colonialization drew new borders and nations. The conquest, genocide, and enslavement of the African continent and the Americas followed. Post-colonial nations around the world became exporters for Europe—aqua-exporters like Brazil.

Europe was first to innovate, conquer, and globalize. Europeans' standard of living improved because of first-mover advantage. That led to aqua-importation, infrastructure for clean water and sanitation, and eventual regulation.

Just like in the US, regulation in Europe made water more expensive. Today, Europe's insatiable appetite for water-intensive aqua-exports diminishes European buying power as water value rises. As a result, its fertility rate has declined, and family sizes are smaller. European families can't afford to *comfortably* live and have two or three children anymore. So families cut back on having children. At the time of writing, the European Union's total fertility rate is 1.6. This number reflects a dying or negative population growth because they are not replacing themselves: 1 female + 1 male = 1.6 children. Why? Based on all their aqua-exports desired and purchased, Europe is shrinking because it wants to afford and spend more on water.

Textbook economists might attribute dying populations in Europe to deflationary pressures because of classical economic benchmarks and indications of slow growth, a.k.a. gross domestic product (GDP). During deflation, economic growth lags and GDP falls. However, it's no coincidence that water inflation also eats away at GDP and income, and thus fertility. What do I mean? Any type of inflation subtracts from nominal GDP. So if France's nominal GDP is 2 percent, while their core inflation is 1.5 percent, and they experience undisclosed water inflation at 1 percent, then GDP equals negative .5 percent, a.k.a. regression.

$$\lozenge > \$$$

Are US populations shrinking as the family size is sacrificed for education and higher income levels? Or is total US affordability lagging behind the growing billions on the planet demanding the same amount of global freshwater being spread evenly amongst everyone? There's a compromising truth. When population growth peaks, will water privatization and commodification be nearly complete, with water prices flashing like gas station lights?

We ask ourselves, are men and women having fewer children in developed economies because of their water diet? Is rising water value causing the price of everything to rise? The answer to both questions is *yes*, and it's called *water inflation*.

Due to rising population trajectories in developing regions, food and energy demands are going to multiply globally. That means having children will become even more expensive. That means our food, clothing, energy—everything—will cost more. When the number one upstream input to all productivity, water, is scarce or highly demanded, the price of everything it makes must increase. And it does. The increase in our paychecks (nominal wages or real wages) has not kept pace with these costs, because governments do not recognize water inflation. Today, fewer children are created in developed nations because families wanted to consume more water via aqua-exports. The irony is that *more children are made where access to water is not available*—and it's no coincidence that where water is scarce, opportunities for education are just as scarce.

According to Waterinfo.org, an American taking a five-minute shower uses more water than a person from an underdeveloped nation uses all day. *One* American on average uses 176 gallons of water daily; the average African *family* uses just five gallons.

One might ask, "Who uses more water per capita—developed nations like America and Germany, or developing ones like Afghanistan and Ghana?" The answer is a resounding "Developed." Which continent makes more babies, Africa or North America? Africa. Yet the lands with lower fertility rates, like the US, use much more water.

Yes, it's a paradox. If a family has more children, then they will buy more house, buying more beds and more sofas and using more showers and toilets. The family will buy more food, take more trips to soccer practice, and buy more fuel, essentially needing more and more water. The more water we need, the more expensive it gets, lowering our desire to have as many kids. In Niger, West Africa, millions upon millions lack water. The *average* family doesn't have beds for every child, doesn't have sofas, doesn't have toilets, doesn't have cars made by water, and doesn't even have enough money to buy aqua-exports like gasoline. But West Africa has the highest fertility rate. Americans use more water and therefore have fewer children, because water is in everything and everything is expensive. If we lived in Niger, we would use very little water. Our fertility rate would be 7.57 children per woman, and, unfortunately, we would hike for water rather than gaining income or an education.

Simply look at the diet of meat-eating, developed nations versus underdeveloped countries like Ethiopia, which marginally maintains sustenance for its people. Developed nations eat water-rich foods like steak, apples, and almonds, and drink water, wine, and beer. More children in the US family means more third-row seats in SUVs and trips for soccer practice. Soccer fields need irrigation, too. The more aqua-exports we buy and continuously maintain—like our cars and the food in our fridges—the more fuel we burn and water we use. Homes, offices, and schools are colossal combinations of aqua-exports. These buildings are made with nothing but water-intensive commodities like concrete, copper, steel, timber, and the oil used for ships to move it and machines to lift it. And developed nations burn more water for fuel to heat these buildings. The more buildings we have, the more water we consume. Water is in everything, and if we consume more of everything, then we consume more water. And that leaves less water for everything else—like raising as many children as we once did.

Source: Belyaev, V., Institute of Geography, U.S.S.R. National Academy of Sciences, Moscow (1987)

Fertility rates are much higher in developing and underdeveloped regions of the world. These developing countries are using less water and having more children. The more water we consume in developed nations, like the US, indirectly forces us to have fewer children.

Kids are expensive, because water is in everything, and very few people are recognizing that invasive, long-term impact on economic realities. If the equation were expressed, it would look like this: Using More Water—Having Fewer Children | Using Less Water—Having More Children.

DEVELOPED COUNTRIES **SPEND MORE WATER** AND **MAKE FEWER BABIES**
VS
DEVELOPING COUNTRIES **SPEND LESS WATER** AND **MAKE MORE BABIES**

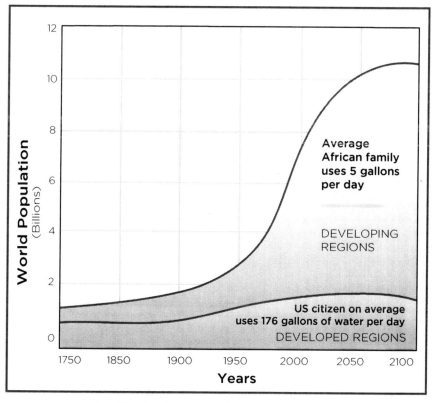

Spend More $ on Aqua-exports which = Costs = Time = Money
Families Shrink as They Chase Their Water Consumption

Ô>$

What if I told you that global gender equality depended on water first? Would that make sense? Social injustice for women is the highest in the world where water is not readily available. For example, "water wars" plague Syria, where women experience undeniable violence.

Giving women contraception in Africa to lower childbirths (quality versus quantity) makes no sense without explaining to women what they could accomplish with an education. How can we inspire them to want an education without freeing them first of their worry about whether they'll have clean water for their thirsty children? How can anyone educate themselves if the fundamental requirement for life is not there and they have to hike miles to go get it? Addressing gender equality, empowering women around the world, and educating children all require water first.

CHAPTER 3

Water Stamps, Before Food Stamps

WHAT IF I TOLD you water inflation was the reason everything was so expensive in California? Would you believe me?

The Golden State sets the stage for anyone's dream. Songs about California evoke images of luxury and vibrate visionary perfection to our ears. California shines a spotlight on the American dream, with the people of Hollywood, its cathedrals of nature, its climate, and, most perfect of all, its food production. Oddly, as we compare it to every other country on the globe, California sounds more like a warning siren despite it all.

Ahead of *entire countries* like Canada, Mexico, Germany, and Spain, California is a leader in agriculture and economy. California feeds the world, and because of its responsibilities, any shift in its food production ripples through the global economy. Hence California is an indicator of what could happen to international food prices because of this state's force.

Commodities worldwide are linked to the supply and follow the prices of California's produce. If avocados suffer from a California drought, the price of avocados sold from Mexico can easily rise.

Assuming Mexico has a below average harvest of avocados that year, then the total global price of avocados would naturally increase.

Essentially, a drought from a smaller global producer of avocados will not change the price structure of world supply. However, when California experiences extreme water issues, its sheer size seems to expose aqua-export markets worldwide.

This land is home to lifestyles of the rich and famous. When we visualize it, we can imagine super-sized California Raisins singing against a backdrop of perfect weather and scenic bliss. But that might be just a claymation fantasy. Cali is also extravagant and high-maintenance regarding water. When we peek under the hood of this powerhouse and look below the surface, the way water is rationed in this area is provocative. Before the residents and nature get control of their aqua, this productive land in California gives first dibs to the rest of the world and hoards not only its local resources but also the entire country's water supply.

The Golden State is a breadwinner for the US. California's productivity is linked directly with US buying power, a.k.a. "Brand USA." It is an area of indicators and an opportunity to recognize how water value affects our wallets. There is a fascination with California's droughts because of how responsible it is for our aqua-exports. If total buying power is contingent upon each country's aqua-export performance, then US affordability and purchasing power are tied to how California performs economically to other nations.

"California is the world's fifth largest supplier of food, a big reason why the state would, if an independent country," be the sixth largest economy in the world, according to Matt Schiavenza in "Economics of Drought." California is the elite, beating most of the world's exporters. One could argue that as California goes, so goes the United States' economy. California dominates in the production of many agricultural products and is the exclusive US producer of almonds, artichokes, dates, figs, olives, persimmons, pistachios, pomegranates, prunes, raisins, and walnuts. The state also produces between 70 and 99 percent of eleven other crops, including grapes. Since few other states have the resources to produce them, these crops are known as "specialty crops" and are very lucrative. In 1997 alone, California's production of fruits, nuts, and vegetables accounted for more than half of the United States' total output.

Despite these spectacular numbers, California robs Peter to pay Paul, a reality that breadwinners face to protect their family assets during many short-term deficits. If California is a breadwinner for the United States, then the "family assets" would be US water and the strength of the dollar, a.k.a. Brand USA buying power. While Peter might be the Colorado River and Paul might be China, altogether the US is making the American dream work with teamwork. "United we stand, divided we fall." Yet in life, no matter the consequences, prosperity is a game with economic rules where scores rank our standard of living.

$$\Diamond > \$$$

California inevitably exhausts its domestic resources as it overdrafts its natural groundwater, consistently sucking dry its rainy-day funds to perform in the economy. Deep beneath this land with perfect weather, there is an open treasure trove of natural aquifers—inherited groundwater from all the ages. Fortunately, there are forecasts ahead of sunny skies; unfortunately, insufficient rain means no opportunity for California to fully replenish its water supply. When the state's "surface water" projects are "dry," farmers "return" to "deep" pumping. This not only depletes rainy-day funds but also digs a deeper hole for our *water deficit*. Despite how climate changes for the Golden State, rain or drought, the show must go on for the sake of the global economy. Behind the scenes, or rather, underneath the surface, agriculture and industry drills deeper, to unsafe levels, and then the world beneath our feet becomes a game of Jenga.

California is in jeopardy of collapse, literally and figuratively. And that collapse would ripple throughout the globe, inflating food prices and exporting more water inflation. With no accountability, overdraft protection, or previous regulation of California's underground water, those who have the innovation, means, and relationships can close off this treasure trove for private use. This is a cordial war between public and private rights. Just as collecting rain is illegal for citizens in some states, in Cali, it's a bargaining chip for corporate water take. Access to California's groundwater carry stories akin to its "Wild West" heritage. According to investigative reporter Mark Arax, "No one knows how many wells have been stuck into the ground, how many pumps are pumping out water . . . The counties aren't required to really even keep that information." While this uncivilized "Gold Rush" is now grabbing the country's attention, the pumping is mostly that of private parties seeking gains to the detriment of everyone else's interest. Just one corporation, Nestlé, is bottling roughly thirty million gallons of *public water* just from California—for a federal permit that costs $524—and selling bottles for exponential profit. According to CBS, Nestlé's bottled water sales were up 8.6 percent in 2017, fueling the private water taker toward more water and more profits.

On a global scale, beyond borders and nations, there are overwhelmingly more privatized water supplies than publicly funded water projects—controlling the most powerful flowing *current-cy*, water. This is just another example of the many lives on life support, hooked up to someone else's water supply.

Regardless of private or public use, overdrafting water in California is a scary truth—a risk surprisingly interconnected with past, present, and future global crises. Just like a run on the banks, there are *systemic risks* and local consequences. But in the Golden State, all water withdrawals are overdrafts. To keep up with global aqua-export production, California and its residents pay higher water inflation, a.k.a. living expenses. Desalination is a proposed solution, but the price increases would be astronomical, yielding more water inflation with strings attached. For example, San Diego would be charged "twice as much" as the traditional water costs paid by the Colorado or the Northern California region, according to *Search for Drinking Water*. These desalination plants are required in armies just for Southern California to meet 30 percent of existing demand, meaning billions and billions of expenses and IOUs. All the while, the plants themselves are extremely energy-intensive, and thus water-intensive, and the ocean environments at risk are a whole separate subject.

◊>$

Behind the glamor of ranking highest among world producers, California continues to send aqua-exports around the globe, exacerbating a water deficit. According to the Pacific Institute, "Half of the water consumed in California goes to produce things the state exports . . . This export of 'virtual water,' as it's often called, essentially means sending water out of the state . . . to places like Japan and the United Arab Emirates rather than elsewhere in California." These countries are largely devoid of food, essentially importing American water through the products they buy from aqua-exporters in California. One of the biggest water crops is alfalfa for cattle. "A hundred billion gallons of water per year is being exported in the form of alfalfa from California," writes Professor Robert Glennon from Arizona College of Law. "It's enough for a year's supply for a million families."

Because of California's immense footprint on the US and world supplies of aqua-exports, this region serves as a water siren, driven by supply and demand. With water in everything, and everything considered an aqua-export, it is essential to understand how water impacts global prices and buying power, with California as the frontrunner.

The Pacific Institute claims that all the water required to make the food, clothing, electronics, and other products that Californians *import* amounts to more than forty-four million acre-feet. That's more water than would flow unimpaired down all the state's major rivers in a year, and more water than would fill all the state's reservoirs. In sum, more than twice as much virtual water comes into California as it exports.

The sacrifice of water needed for California to maintain global economic dominance has to be expressed somewhere, especially when the water *imported* covers a shortage for production in the state. The economy grows, and the people of California pay higher costs because water resources that the state requires are earmarked for the whole country. Essentially, the people of California are taking one for the team. They're doing it for Brand USA and its GDP score. These citizens experience more water inflation and, as a result, have less buying power. Eventually, all of us will pay higher costs to maintain every breadwinning state's never-ending water demands and deficits.

US currency buys less in California for two reasons—because it needs more water than it has for actual manufacturing (water deficit), and it taxes the highest environmental regulation among its fellow states. This tax is an act to recognize water value, a PE (positive externality) for the state's environment and people. The higher prices and standard of living in California pay for the imported water necessary for California to be one of the largest global economies—and with less smog than decades past. California's water inflation is expressed as a loss of purchasing power by state residents, simply revealed by higher-priced goods and services, and a volatile and regressive return to homeowners. This will be addressed in later chapters.

When we see water as an aqua-export calculated in GDP, basic economic principles declare that if a country or region the size of California buys (imports) more water than it sells (exports), the brand of that country's currency eventually becomes weaker. As part of the GDP, the balance of trade equals the sum of all exports minus the sum of all imports. The higher the value of imports entering a country compared to the value of exports, the more negative that country's balance of trade becomes. Imports subtract from any nation's GDP. California may outperform such nations as Greece, but if it were its own country, it would be approximately the sixth largest welfare nation.

It has a deficit economy, subsidized by other states' water, because after all, the Colorado River is an "entitlement" among seven states.

Investopedia calls a trade deficit an economic measure of a negative balance of trade in which a country's imports exceed its exports. With water, California has such a deficit, leaving its GDP and currency—and, hence, its buying power—vulnerable to outside forces, such as inflation of various kinds. This triggers a reduction in buying power, meaning prices push higher. Eventually, any inflation leads to interest-rate decisions by the "coaches" at the Federal Reserve (known as *monetary policy*)—financial adjustments with consequences for the US team. Investopedia also notes that "a trade deficit represents an outflow of domestic currency to foreign markets" to purchase imports. In the case of California, we mean buying and importing water from other states. Just like Greek banks in 2015 were running out of currency and experiencing social upheaval, California's economy is on life support regarding water for aqua-exports, and they pay for this risk through higher water inflation. That is a value that has no accepted space in GDP calculations, but it is the elephant in the room. That is the aqua-export-GDP-currency relationship at work.

We usually do not think of California as bankrupt or debt-ridden like Greece. But we can now see how they are similar. However, the world impulsively still counts on the US. The only significant difference between the United States and the European Union (EU) is fiscal federalism—the "ability to transfer economic resources from members with healthy economies to those suffering economic setbacks," as Pulitzer Prize-winning economist Paul Krugman and colleague Maurice Obstfeld observed. In the US, "states faring poorly relative to the rest of the nation automatically receive support from Washington in the form of welfare benefits and other federal transfer payments that ultimately come out of the taxes other states pay." The ability to sell a US bond is fiscal federalism. The debt sold from Treasury bills supports all US states. The US government uses the rules of fiscal federalism to its best advantage in a managed perception of the country's resources and Brand USA.

California would likely be closer to insolvency without the welfare of other states' water supplies. In the United States, the bankruptcy and debt caused by water shortage of any state can be spread evenly over all other states with sufficient water supplies by way of US bond sales. Fiscal federalism creates a sovereignty for national bond sales so that the US Treasury can average all the states' risks into one vehicle: a US Treasury bond. So if California is short water, but Minnesota or Wisconsin is not, the country's debt may be minimized because the risk of investing in the US through a bond is spread out evenly through all the states and their productive capacities. The US Treasury can then use the proceeds from selling the bonds to prop up deficit states like California.

Considering that California accounts for 14 percent of US GDP and 3 percent of total world GDP, there must be a consequence for the US dollar if the state is short water for production. And it's not the only "water deficit" giant in the United States contributing to massive *water gaps*. On the southern border is another giant aquatic sponge—Texas.

$$\Diamond > \$$$

As we introduce thirsty Texas, we become aware of the war over water between agriculture and energy demands, and just how exactly a water stamp becomes a food stamp.

The Lone Star State is the largest-grossing producer of oil and natural gas in the US; Alaska is next, while California sits just shy in third for oil production. Texas may only need seven barrels of freshwater to procure oil, but it takes 10.5 barrels of water to make that same barrel of crude in California. While we are talking about aqua-export commonality, Texas also unloaded 7.4 billion barrels of produced water onto the state from oil and gas wells. "The old-timers called it drip gas. You run out of gas out in the desert bringing your herd of sheep back, you'd fill your truck up with drip gas and drive on to town. That way you don't have to walk. It'd smoke like hell and make your motor ping, but it'd run" (*Split Estate*). To underscore the massive NEs (negative externalities) and ULs (unrecognized leaks) inside Texas, consider that in 2007, oil and gas generated 35 percent of the total produced water in the US, oversupplying its borders with a hazardous waste stream bearing the highest volume of produced water of any state. What this all means is that Texas must have the most questionable water quality of any state. We don't just put drip gas in our tanks in California, it seems; it helps grow our food as well.

Evidence of questionable food quality and future higher food prices, oil prices, and eventually water prices are better understood when we consider that the Texas oil and gas (O&G) explorers are wrangling the farmers of the state for their water supply during growing seasons. The water wars are taking place between explorers and farmers in boardrooms during business meetings.

What happens to the price of aqua-exports—electricity, fuel, and grocery bills—when water gives us a 100 percent effort, but the supply of it still falls short of our demands?

In 2008, when the first well was drilled in the Eagle Ford Shale, Texas O&G was not openly competing against farmers (and agriculture) for water supplies. Yet in 2011, food was being challenged and its traditional supply of water chastised and questioned by energy's competition for its use. While some farmers traded water, others declined. Due to the hydraulic fracturing boom, Texas became home to the biggest natural gas production sites in the US. Hydraulic fracturing is also known as fracking. The petroleum industry first employed fracking in 1947. Halliburton Oil was commercially licensed and operational with such processes in 1949. Hydraulic fracturing involves shooting powerful cocktails of water infused with hundreds of chemicals and sand into underground wells, encouraging O&G to flow liberally to the surface. Brodrick poetically explained, "Water injection increases the flow of oil, so as oil fields deplete, they need more water to keep production steady." Quantitative analysis may well conclude that oil tycoons are water junkies, because they are addicted to oil, and therefore to water.

On Eagle Ford's shale formation, according to Bloomberg, fracking one well "requires as much as thirteen million gallons of water, enough to supply the cooking, washing and drinking needs of 240 adults for an entire year."

During a raging drought where 94 percent of Texas was severely short of water, oil and gas companies had to take water for production from aquifers, farmers, irrigation districts, and municipalities—scraping up just enough to get by, even so early in fracking's boom.

Hidalgo Irrigation District No. 2, located in San Juan, Texas, distributes water to farmers pursuing four hundred thousand acres of cantaloupe, cotton, peppers, and sugar cane. The district was selling water to the oil explorers during the worst drought since records were kept. General Manager Sonny Hinojosa was not sure he could keep selling water to the oil companies if rain did not arrive and replenish the reservoirs within a handful of months.

Forecasts estimate that drilling will "explode" over twenty-five years in Eagle Ford. The Texas Water Development Board and the Bureau of Economic Geology from the University of Texas feel that hydraulic fracturing would multiply water use by 1,000 percent by 2020 and 2,000 percent by 2030. Hence, we expect water to "give 140 percent."

When oil companies began approaching farmer Bruce Frasier of Carrizo Springs, Texas for his water, his family has been farming forty miles northeast of the Rio Grande for 100 years. Frasier denied water sales to the O&G sector, which offered a mere $0.40 to $0.70 per barrel for his crop's water (a barrel equals forty-two gallons). The water shortage caused the loss of more than half of Frasier's cotton crop and a reduction of 70 percent of his cattle herd.

Before the drilling of the first Eagle Ford well, there was no market for a farmer's water in the area. But in other states, farmers get in line as though they are waiting for a Super Bowl to convert their water rights for exploration and production companies. Exploration and production companies are creating competitive market demand for water rights, which increases the sacrifice for water use, capitalizing on demand and fueling the fire for a new and separate water market. Prices for all aqua-exports would rise under these conditions. If they don't, then water value remains unrecognized, and future price shocks for food and gas swim like hungry sharks.

Considering fracking's water footprint in producing O&G—and fracking's proprietary damage to freshwater—reveals fossil fuels as easily our most insincere form of energy. The fields where drilling, exploration, and gas well production take place are called "shale plays." The two most productive shale plays in the US are in Texas, with the top one in Haynesville. Retired government worker, geoscientist, and author David Hughes knows something about these shale plays. He worked thirty-two years at the Geological Survey of Canada. After nearly forty years of energy research, Hughes responded to the hype around energy independence and published *Drill, Baby, Drill*. He analyzed the costs and productivity of many individual wells from both major and minor US shale plays. Regarding shale energy production meeting US energy expectations and requirements, he said, "Once you get on to the shale bandwagon, you're onto the drilling treadmill."

Mr. Hughes talked and wrote about the discovery of what some call the "Red Queen Syndrome." So let's meet the Red Queen. She is like beginner's luck. The first month's payoff from an oil and gas well produces a one-hit wonder of fossil fuels—following that, productivity drastically and forever declines from that peak. But when we burn fossil fuels, we leave behind this permanent produced water—something pure, now changed to something hazardous. Economists simply say that nothing tastes as good as the first bite. The Red Queen and one-hit wonders know this all too well. Mr. Hughes referred to this effect when Haynesville recorded a 68 percent decline in production from the first year of gas recovery. The second year's decline rate was 49 percent, the third year's 50 percent, the fourth year's 48 percent. Since Haynesville was discovered in 2008, billions of dollars have been poured into new wells to offset the decline.

According to Hughes in a CBS interview, that's roughly $8 billion a year spent on drilling just to maintain production. Haynesville has to replace 52 percent of its gas production each year "by more drilling just to keep production flat." If Haynesville stopped drilling new wells, production would fall by 52 percent, he concludes. That's not very sustainable in the long term, nor is it a solution. Altogether, "tight oil and shale gas require about 8,600 wells per year at the cost of over $48 billion to offset declines," Hughes stated. Essentially, shale gas, which represents 40 percent of US natural gas, has plateaued since December 2011. The O&G industry's excessive water demands and pollution are for short-term gains in production.

So, if we're exhausted by our time on the "treadmill," then realistically, once we stop running, we go backwards and regress.

All of this points to one unassailable fact: in terms of water—"energy's energy"—and sustainability for unconventional drilling, the US is in the red. The process of finite energy extraction is on a treadmill. Far worse: it's not just paper money we are burning—we are burning out what constitutes freshwater. The debate on qualifying the massive needs of water for the O&G sector portend a less than "clean energy." The fuel for our bodies, the nature of our blood, becomes shortchanged and contaminated for non-renewable energy. Because of its false price structure, "clean energy" continues to compromise all the many invented, renewable alternatives and much cleaner substitutes.

Just as California receives actual physical water subsidies from other US states, drought-stricken Texas gets something else. Rather, Texas enjoys 2.9 times more government money than California from the US Department of Agriculture, according to *California Agriculture: Feeding the Future 2003*. Since water is naturally scarce in Texas and it puts so many food-and-energy draws on the supply, the subsidy of currency from the USDA offsets the water deficit with digital dollars funneled into the state's bank account. While California gets a water subsidy or *water stamp* from Colorado, Texas gets the almighty substitute—"digital dollars," called *currency*. Printing digital dollars is the only possible way to print enough water or to paper over the shortage. Either way, it's all a water stamp.

Texas allocates this money wherever water deficits create debt, known here as the *water gap*. Texas may look like an empty waterhole in a dust storm, but the water gap is figuratively expressed in society. It is easy to imagine the downstream impact of these water gaps in Texas: the shortage of funds for new bridges, pipes, and roads; the increased healthcare dollars spent due to produced water drying up on dusty roads, causing increased respiratory diseases; underfunded police departments leading to spikes in crime; or less money for all forms of public schools. The water gap is an insidious drain on resources and our ultimate standard of living.

The Federal Reserve already acknowledges this phenomenon in education: "Undergraduate tuition at four-year institutions . . . has risen since 1983 at a rate more than twice as fast as the overall rate of inflation." We can certainly suggest that water inflation is the difference in phantom gaps in government measurements of inflation.

To be the greatest producing states within the United States, mega-producers like California and Texas either need physical water entitlements (like the Colorado River) or digital dollar subsidies (like USDA money) to afford water for operations. Simply put, "Made in the USA" is running an economy with a constantly growing water deficit. The US overdrafts its water and prints currency for the cause— its water needs. This history admits that states need *water stamps* for private enterprise just as much as people depend on *food stamps* for existence. The pattern we must recognize is that when we start off with water stamps, we end up with food stamps.

Can we imagine water inflation eating away at US budgets, as so many programs for entitlements, subsidies, and welfare expand each year? These federal programs are treating the symptom of water shortages. We're simply covering up water deficits and printing dollars to dismiss water's limits. When this house-of-cards strategy collapses, we'll be victims who built our own paper rafts.

$\bigcirc$>$

Just like California, Texas, or Greece, the United States imports and uses more water—in goods—then it exports. And the US, too, has a trade deficit that leaves our currency and hence buying power vulnerable to outside forces. That harms the fundamental health of the country and its GDP score. This creates inflation, changes interest rates, and leads to more liabilities—debts for the people. So as the US operates with a trade deficit, it sells more Treasury bonds (T-bills) and expands the federal budget levels to afford more government spending, giving our players tools to prop up our short-term buying power in the world to maintain rank and strength for Brand USA, the dollar. The government selling T-Bills creates obligations and interest payments to countries like China, Japan, and Saudi Arabia. The more T-bills we sell, the more the US dollar *weakens*—as those other nations take more control of US assets and the perception of US debt *grows*. Meanwhile, debt increases because the US consumes or spends (imports) more money than it makes, saves, or sells (exports) for profit.

The greatest tool for the US Government is time. Using T-bills to buy more time is a technique it has mastered and perpetually relies on. However, the strategic moves of government are made in different time zones from the consumers. We the people are vulnerable and jet-lagged. Boiled down, printing money to offset water's deficit now appears to be normal behavior. The government's tools, more specifically, are to:

- Redirect undervalued water from Colorado to Southern California (entitlement)
- Place digital dollars in state bank accounts (water stamps)
- Provide crop insurance for desert-like Texas by the USDA (subsidy)

- Organize operations of the Federal Emergency Management Agency (FEMA), to prevent a long-term collapse in home prices, whether too much or too little water is affecting the market (market failure)
- Use many other forms of dependency and intervention as well

Because water deficits are pervasive in the US economy, water inflation is constantly working against the American people. That is a danger to our lifestyles and the middle class itself, because any inflation reduces our buying power, meaning prices eventually push higher relative to the past, but our paychecks do not. If the middle class is shrinking, the lower class is growing, and with water inflation, no one is calling the US government to account for this peril.

Our expected standard of living is rising at an undisclosed rate due to water. We're paying more and getting less because water value is unrecognized. Why is the real cost of water use fragmented or the real quality of our groundwater not accounted for? What benefits did we gain and lose by omitting this game changer? If the lifestyles we can afford are based on how much cash, credit, and savings are in our accounts, then we should also be able to hold our country and its water piggy banks accountable.

California or Texas may be at risk of defaulting on debt, but they cannot default because other states support them. The big production of California robbing Peter to pay Paul, propping up the brand of the US dollar, and the government—with its corporate partners—using tools to creatively prop up our buying power, *are just temporary fixes*. It is inevitable for everyone to experience higher costs of living, known here as water inflation—and some states sooner than later. While Brand USA appears a powerhouse compared to the rest of the world's economies, what will happen if our breadwinners, California or even Texas, get hit with a "mega-drought" and can no longer rob Peter to pay Paul?

We need to acknowledge that the more water reserves a country has, the more options and opportunities it has to maintain a powerful scorecard for global economic rankings and financial freedom. The competitive growth and productive needs of any state or country depend on water, one way or another. Conversely, a world with fewer water reserves, fewer options and opportunities, creates more liabilities, debts, and deficits, and leads to some form of dependency, entitlement, and subsidy.

Once we all buy into the fact that water value is a game changer, then we can continue achieving the American dream. We must allow water value to appreciate on our terms, under our vision, or the future will be all pain, no gain.

CHAPTER 4

Made in the USA

THE AMERICAN DREAM ONCE included "Made in the USA," but hasn't so much since the 1970s. That is when our corporate partners officially became multinational corporations (MNCs), moving manufacturing to other countries. Legend has it that "Made in the USA" lost out to cheap labor markets. There's some truth to that. All the while, corporations were dreaming in color as they quietly got in line for new sources of freshwater the American people knew nothing about. MNCs packed up and "outsourced" their need for overpowering water supplies with little to no environmental regulations—eventually to settle in South American nations and Asian tigers like Thailand and Vietnam today.

Vietnam's Water Supply

Water became more expensive in the US after environmental regulations such as the Clean Air Act of 1970, the Clean Water Act of 1972, and the Safe Drinking Water Act of 1974. And the precursor to these laws was the Federal Water Pollution Control Act of 1948. Before regulation, US-based companies grew organically, using cheap and abundant water as their primary raw material. And before federal law, they discharged effluent and polluted water, with little or no cost.

Classic examples were the automakers and material manufacturers in the US—cities near big bodies of water, like Detroit and Pittsburgh. They were North America's China prior to the 1970s. They were aqua-exporters. Remember the forty thousand gallons of freshwater to manufacture a car and sixty thousand gallons to fabricate a ton of steel? EPA regulations made *all* this expensive. The car industry (Detroit) might blame unions, while the steel industry (Pittsburgh and West Virginia) could blame trade policy. But the real cause of their demise was increasing the costs of using and treating water. Back then, no one connected the dots and told the American people that the value of water had changed. When the EPA came on the scene in the 1970s, it gave water a higher value, but never explained that higher value to the American people. Rather, the EPA slapped domestic manufacturing on the wrist for polluting and became misunderstood in the process.

Throughout the twentieth century in the US, water and wastewater systems constantly needed upgrades for efficiency and to improve water quality diminished by industrial growth. Factories and towns needed freshwater to operate and a place to dump polluted discharges. With the attention on US-centric environments, the cost of water and wastewater processes began to climb considerably. Industrial farming was forced to be more selective with choice fertilizers and reduce exposure to specific chemical runoff—the type of chemicals that make crops invincible but the land and people victims.

The Clean Water Act of 1972 and the Safe Drinking Water Act of 1974 did more than establish the "basic structure for regulating discharges of pollutants" and monitoring surface waters while protecting our drinking supplies. These acts determined that many US companies heavily reliant on water supplies could not continue to expand, or at least not as profitably. Simply put, manufacturing became a lot more expensive. Costs began to eat away at well-paying manufacturing jobs and the middle class that relied on them. Water costs were rising, making the US less competitive. Unbeknownst to the American people, water value was rising. White-collar services were demanded, and household investments in higher education moved up with water value in lockstep.

As more companies moved manufacturing overseas, the US became a net importer of aqua-exports. That trend, which gained momentum in the 1970s and beyond, created the trade deficit we now hear so much about. And that deficit was changing the job market, eating away at the value of our currency and the American dream itself.

The middle class shrank in the United States over this era as water's force was hidden and people's purchasing power confiscated through a weaker currency—a weaker Brand USA. The US stopped making things and started promoting less water-intensive services. The United States was no longer a net aqua-exporter. The American flag, which waved effortlessly from a child's hand on the Fourth of July, was no longer "Made in the USA." The auto industry in Detroit and the steel industry in Pittsburgh faded, while the dream remained in American expectations.

The US became a service-based economy in 1975 and began importing more water-intensive goods than it was exporting. Its trade deficit was an attention-getter. The US created more earnings and revenues for other countries as it gave these nations dollars for goods they produced. The exchange was a flood of aqua-imports into North America and a flood of US dollars out, all around the globe. This created a surplus of US dollars internationally, and that drove down demand for the US currency. The United States wound up losing jobs and had less buying power because its currency was oversupplied to the world.

Despite all the efforts to import cheap, unregulated water, anything with water in it became more expensive. The cost and price of commodities climbed independently of US currency and more dependently on the expression of rising water value. National debts escalated. Connected to that was a loss of purchasing power by US citizens. As water grew in value because of growing populations and regulations, the US dollar weakened. The US was experiencing a new, permanent trade deficit of aqua-exports, which created a disappointing GDP. Water value exposed itself globally during the oil crises in 1973 and 1979. All the wars started to have a complex texture that many people couldn't quite feel or see. While gasoline had a smell, water had no odor. With water value in mind, can we make this connection to the oil crisis now?

At the same time the EPA gained traction, the gold standard ended. The US dollar was collapsing in response to something besides the obvious shiny excuse—the fundamental shift in unrecognized water value. We had what economists call a "floating" currency, based on the relative value of the US dollar versus other currencies. Before the world adopted *fiat money* based solely on faith in the 1970s, the gold standard linked the value of US dollars to some fractional reserve of physical gold in banks. For thousands of years, gold and silver were part of a barter system, along with rare spices and food-preserving salts. After the invention of fiat money, governments did not want citizens to own precious metals. Countries instead wanted their people to "drink the Kool-Aid" of their nations' brand and give up the relic and intrinsic value of gold.

When the shiny metal stopped speaking for the US dollar, the "next man up" to replace gold was more connected to every country's currency than ever before. It was water. Because we use it for everything, its abundance and availability are linked directly and indirectly to the way nations value currency. Countries with unregulated water supplies could make aqua-exports cheaply. Nations with environmental regulations had expensive sources of water and were at a *comparative disadvantage*.

As we expand in population, the middle class will continue to weaken because of the global rise in water value. It may grow in some areas where a middle class is emerging, but regress overall as nations become more developed and regulated, exacerbating a global water shortage. What's needed is a system that more closely ties the value of our currency and purchasing power to real water costs. When water is priced according to its true availability, only then will we be able to recapture the competitive edge. "On average, every US dollar invested in water and sanitation provides an economic return of eight US dollars," according to Water.org. Who would have thought water could deliver so much in profits to our economy?

If we could make 800 percent on an investment, would we join the other lucky investors in line to secure those gains? What I think when I read this is that if we invest in water, we would have the money to improve water systems *and* use it more efficiently *and* return exponential purchasing power to the people. It would no longer be possible to waste 30 percent of the water we consume through leaky pipes (non-revenue water). We would subconsciously conserve water and, by doing so, have a lot more of it to make things. With abundant and reliable supplies, we could again become an aqua-exporter, creating jobs here instead of watching them move elsewhere. We could be water efficiency gurus— "water efficienados"—making the industry sexy.

◊>$

I mentioned the gold standard above for both historical and illustrative reasons. Gold is a commodity. It has a value. It's a hard asset with a real price that reflects demand for it.

Gold will always have an increasing value. Why? Because of the *paradox of value*, also known as the *diamond-water paradox*, an accepted model established by Adam Smith in *The Wealth of Nations*. Gold is scarce, like diamonds, with high value in exchange for the most abundant resources on the earth—which, at the time of Smith's findings in 1776, was abundant water for many populations. Since then, gold increased in price with demand, as populations surged. At the dawn of the 1970s, the masses were perplexed to be having a conversation about water's value, whereas today, it's a mainstream "commodity," considered "precious." Can we say gold and silver are now a representation of water, especially since the early 1970s? Yes, we can.

Precious metals will have buying power in the breadbasket of a country like the United States; however, the only value of gold in the desert is for a water trade. Gold means nothing without water. Empires, kingdoms, and nations used gold and silver for thousands of years to transfer a store of value, and as a medium of exchange for goods and services.

Earlier civilizations had sufficient water supplies relative to their nearby pockets of population, which obviously demanded water resources. Historically, it was gold and silver that were more finite and scarce, just as clean water is today for so much of the world—gold relative to the people in history, and freshwater relative to the people today. When gold screams higher in prices, it is saying something substantial about aqua-exports. Gold is the translator for water value today.

In 2011, Federal Reserve Chairman Ben Bernanke made a statement for central banks around the world. Bernanke claimed that gold is not currency nor money, yet central banks own trillions of the precious metal. Controlling the money supply reduces real purchasing power in many ways—like a rigged system. The trends in history have demonstrated that the behavior of gold prices marched higher every time it was confiscated or marginalized, as with Executive Order 6102. The more governments that decouple the relationship between gold and their currencies, the higher gold climbs in value over time.

Gold is a force against currency, and owning this tangible force empowers the holder. When global citizens turn their head to fiscal policy by government and dislike monetary policy by central banks, they sell their currency and purchase gold, as if protesting the government and its financial choices. Other people sell their currency for the same reasons and buy art, another tangible item with a means to preserve their financial freedom. When the middle class maintains its buying power, it gains more voice, and all economies benefit, but if the public has less buying power over time, it will have less money, less influence, and thus less power or voice. Less of a middle class equals fewer voices from the public.

Head coach Ben Bernanke was correct. Gold is not a currency; it is a *current-cy*—a force pushing against the government and a voice speaking for the people. Gold holds governments accountable for printing money or shelling out more digital dollars, and moves higher over time. Many gold purchases take place free of taxation in the *informal economy*, otherwise known as the *black market*. Yet fiat money won, as international commerce and trade over the course of history found beauty in digital currency. After all, digital bank accounts are measurable, are very light for travel, and can be tracked, refunded, and taxed. And the margins between what a seller wants and what a buyer will pay are less far apart when exchanging currencies than compared to gold. Regardless of gold's setbacks, every time a citizen buys gold and sells currency, they are telling their country that they don't like the options offered by their brand.

Some economists and anti-fiat marketeers want to pin the responsibility of inflation on the decoupling of gold and silver from the world's reserve currency. They are not wrong; they are just focused on the wrong aqua-export at the wrong time. Gold was the rage once, but it is today just a translator of the global populations' true demands: water. Regardless of currency, water's value was always stored in gold and silver.

With billions of humans on this planet now demanding exponential aqua-exports, precious metals mean nothing unless we recognize water value. Printing *more* fiat money lowers the buying power of that money and increases the price of aqua-exports like gold. It's a smoke-and-mirrors excuse for why commodities increase in price. Printing money drives up the value of everything that is water-intensive—everything but water—and so the act of printing money actually confesses that water has no alibi. It proves water's force, because if central banks did not have to devalue currencies, they would not, but they have to speak for the shortage of water. Central banks print money because dollars disappear to pay for the higher costs related to water.

$$\Diamond > \$$$

The currency market, or foreign exchange market, is the most liquid financial market in the world. Today, more money transacts on this market daily than any other flow on the globe—an estimated $2 trillion a day. Currencies change hands during international trading twenty-three hours a day.

These massive currency markets emulate the characteristics of water. Currencies ebb and flow like the tide, move violently like the weather, flow in trading like a current, and literally and figuratively float the value of physical commodities, like aqua-exports. Despite its bears and skeptics, the US dollar is still the greatest and ranked the most liquid currency. Nearly every central bank and institutional investor in the world holds US dollars, and some foreign countries use it as an alternative to their local currencies. Because of such stature among all other currencies, the US dollar has invasive and substantial power—just like water. Hence the US dollar and US government have a responsibility to the free world not to export inflation. Since the 1970s, money printers around the world responded to new US water regulations and higher water value, especially with the US currency in so many international banks. Water inflation appeared in the increased value of commodity prices purchased with dollars rather than in the US dollar's depreciation alone. World populations expanded, demanded, and squeezed water everywhere, and the EPA's regulation catalyzed unrecognized water value—developing a water gap in US money for all holders.

Meanwhile, US imports exceeded exports, spending exceeded savings, expenses exceeded income, and debt accrued in the United States. Because back then and today, there is no outlet to express water value, water has no voice, no currency, and no buying power. So guess who loses?

To finance that mystery debt for citizens and corporations, paper money is printed, and this act of welfare causes the currency to weaken over time, resulting in a falling buying power, or the cost of aqua-exports continually rising. That shrinks the middle class. Even with the rising value and strength of the US dollar between 2015 and 2017, have fruits and vegetables' prices fallen at grocery stores or restaurants? That's how we know that water has gone missing from our wallets.

Central banks discretely perpetuate a message that freshly printed money and cheap money—low interest rates—can stimulate growth and be responsible for "real" rising inflation. This is not so true. Since the 1970s, central banks, like the Fed, could have in fact printed money to excuse inflation, like a scapegoat. This might not be intentional, but the fact remains: they are printing water, not money. New paper money creates inflation and covers up the lack of cheap freshwater available to run all economies.

Despite this profound relationship, central banks have no control over water's force. After the 1970s, money printing has simply been a translator for real value—*water value*. Nations need stable levels of inflation—and deflation—to grow. They also need a stable water supply. A stable water supply is just like the weather, as consistent sunny days benefit sales at stores and consistent blizzards and hurricanes wreak havoc on the short-term economy. How can we ask the earth's water supply to be stable? Nations don't ask water anything; they print money instead. Going through a drought? No worries! We'll print money and lower interest rates to spur borrowing, despite the rising prices of everything else.

Like the wild movements in the price of gold, water is a *current-cy* because it is a catalyst and a store of value for wild movements in the price of raw materials. When the price of raw materials like commodities are out of control, governments have to acknowledge and help the public digest these changes in affordability. Governments use tools like crop insurance and subsidies, and inflation numbers coming from the CPI and Personal Consumption Expenditures get politicized. As the Federal Reserve Bank of Richmond said, "Inflation is a concern of policymakers for good reason: it eats into the purchasing power of households' money and effectively taxes their consumption of goods and services, even if they are not always aware of it."

Logistically, water cannot be moved and stored effectively, because it is very heavy and voluminous. Hence, *current-cy* is something cumbersome and heavy in value. The same is true for gold, which is very cost-prohibitive to store or transport because of its volume and weight. While gold is still exchanged as a substitute for fiat money, it is difficult to sell.

Unlike gold, water has never been stored at central banks. Governments have never priced it transparently, nor has the black market. But like controlling gold, controlling the water supply reduces real purchasing power in many ways—like a rigged system. The behavior of aqua-exports has marched higher every time economic benchmarks like the CPI marginalized water.

Meanwhile, purchasing water rights is an admission of water value and a hedged investment to parlay governments who push their paper money—IOUs. With hindsight, the more governments decouple the relationship between water and inflation, the higher aqua-exports like gold and meat climb in value, and buying up water rights seems much more logical to those who cannot see the better picture. If we valued water properly, then more of it should be available to the public and held by the public, since residents pay the most for it. That empowers all consumers, not just the few who own water rights. And those consumers go out into the economy and support corporations with their buying power. If we fortify the population through water, how strong would US GDP be? Citizens and corporations alike would thrive. If we squander water and ultimately have less of it, it becomes even more precious and therefore scarce. Countries do not grow without their peoples' participation.

Any rigged system has a process respected and understood by few. Sean Brodrick acted as the resource strategist for the Oxford Club and commented about rigged markets in the discussion of silver's fixed history:

- The gold fix was rigged, according to Britain's Financial Conduct Authority, which fined mega-bank Barclays over a trader influencing the gold fix in 2012. Today, twenty-seven lawsuits related to the gold fix are winding their way through US federal courts.

- International oil prices were probably manipulated. European authorities raided the offices of BP, Statoil, Royal Dutch Shell, and energy news publisher Platts over that.

- In fact, since 2008, regulators have uncovered price-rigging in everything from currencies to interbank loans. Rigging in that market went on from 1991 to 2012.

When the system is rigged, we don't have to pay attention to what's causing it—we just point fingers. The events above are all a cover-up to treat the symptoms of water value rising globally. Like treating symptoms to cure disease, international *currency wars* through central banks' monetary policies are really *current-cy wars*.

Currency wars occur when one nation devalues its currency against others to make its exports cheaper for other countries to buy and therefore more enticing and easier to sell. China is persistently accused of keeping its currency weaker to do just that. Currency wars create higher demand for aqua-exports from a country, which are valued relatively lower through currency (and relative to other countries' currencies). These aqua-exports are sold below their intrinsic value—a false demand, and a practice which ignores water value, and thus instigates more water inflation over time. That's a moral hazard for us all. When major economies like China, Europe, or Japan devalue their currency, they drive up the value of the dollar, regardless of the true health of US economic fundamentals and rankings such as GDP, inflation, or interest rates. Why? Because Chinese, European, and Japanese central banks sell their currencies and buy dollars; therefore, the value of those currencies fall and the US dollar strengthens.

Breadwinner China has 22 percent of the world's population and 9 percent of its arable land, *but only 6 percent of world water*, according to *The Brock Report*. This means that when the central bank in China prints money, it is attempting to *print water for its population*—to cover up that gap and offset the impact of water shortage in China. Countries do this simultaneously to deal with their very real water shortages. China uses the profits from its cheap export sales—created through currency manipulations and currency wars—to buy water from California by importing alfalfa grown in the Golden State. Full circle, when the People's Bank of China sells the renminbi to keep its currency much lower in value, it is either buying another country's currency or their debt, like US dollars or Treasury bonds. And the world's reserve currency, the US dollar, benefits whenever this happens.

Currency wars are really a race to print the mirage of sufficient water supplies through digital money for the people.

Nobel Prize-winning economist and Princeton University Professor Paul Krugman says, *"Seigniorage* is the name economists give to the real resources a government earns when it prints money that it spends on goods and services." Central banks cannot print *current-cy*, so they print *currency* to cover up the growing debts created by their unrecognized water gap. *Non-violent water wars* are how China affords to buy water elsewhere in the world: through aqua-exports and aqua-exporters. We will hear of China in Africa and Brazil, buying vessels of commodities and actual producers of aqua-exports, and we will know China is doing so to close its water gap. In the end, we will hear that China is experiencing high levels of inflation, and we will know it's because of its *very real* water inflation.

While some governments have the tools to mask their inflation, others do not. "Real resources" like freshwater are finite, but paper money is not. The race to print money is not a currency war—it's a *current-cy war*, papered over and propped up with digital dollars.

Central banks will continue to print money to cover up the shortage of water for production means, masking the higher costs of environmental regulations and the impact on their economies. The greatest moral hazard and systemic risk is that of slowly removing water from the equation, despite its importance. As if we are turning our piggy banks upside down and shaking them until we hear nothing, the message echoes down into silence, as we realize water went missing from money.

"See, Tammy? The news is talking about currency wars from China. Remember from *Waconomics*! China has 22% of the world's people but only 6% of the world's water. China races to print money because it doesn't have enough water for its people."

"So when they print Chinese dollars, China is printing money to buy ALL that water needed to make oil in Brazil. It's the same story when China buys meat from Brazil . . . they don't have enough water for their cows either."

"You mean, when China prints money to buy oil from Brazil, the Chinese are buying the Amazon water it took to make that oil?"

"That's right, Tammy! And the news says it's a currency war between central banks when it's a water war among nations that don't have enough water for their people and production."

CHAPTER 5

Water Went Missing
from Money⬤

IF WE HAD TO explain to a child why it seems so expensive or even unaffordable to go to the grocery store today, how would we? After all, we said the same thing about going out to eat at restaurants last night. The same thing could be said about what has happened to *quality*. Quality items are undoubtedly getting more expensive. The items that used to be quality at the prices we still pay end up rusting overnight and being thrown out not that many days later. Is this our buying power at work? Can we easily describe how there's some level of inflation in our lives?

The TV and some books tell us that when we print money, we get inflation. What we've learned thus far is that when we print money, we do it because we are short water for people and manufacturing. We also print money to cover up the cost of growing in our environment and cleaning it up. So printing money becomes a scapegoat for inflation when the real elephant in the room is water. Instead of pointing fingers, let's ask some serious questions. It's time to interrogate the robots.

The Consumer Price Index, the CPI, is the most relied-upon gauge of inflation. It measures the average change in prices from one period of time to another that *urban* consumers pay for a basket of goods and services.

Imagine the CPI as a robot—a powerful, multi-functioning robot that calculates components of the economy and outputs changes in prices. Over time, this robot reflects a change in the perception of our buying power. So what components of the economy are entered into the CPI robot? Who is responsible for this data? And has the robot included and recognized water value the same way the EPA did in the 1970s?

$$ \bigcirc > \$ $$

This is a layered conversation that involves how the government measures price change and why our wallets don't keep up. It's a world largely ignored by the mainstream media and the general public, and not even fully understood by many bureaucrats. I'll do my best to keep it entertaining, but be aware that this requires attention. If you could fully understand why your money is leaking simply by reading through a few layered conversations, would you?

A team of economists, financial folks, and mathematicians at the Bureau of Labor Statistics (BLS) chooses and enters the data. Imagine the BLS staff in simple terms, as if they are college professors who calculate grades on a weighted curve. Like the grades, each component of the CPI is weighted. The CPI robot analyzes components like trends or popularity contests of staple goods and services we buy and how their prices change. As we've learned, all goods and services end up being water-intensive users and therefore we can consider them aqua-exports. So why the big production? Because the robot reflects a gap. Economists and financial advisors use this gap to determine the difference between what we paid today and what we'll have to pay tomorrow to plan for financial freedom and retirement. We use robotic calculators such as the CPI as a basis to stretch our earnings today and to beat inflation tomorrow.

But what would happen if the college professor used a robot that overlooked the most important details when considering the curve? Or what would happen if the professor omitted the highest grade from the curve? Either most of the class would pass or most of the class would fail based on that grade. As it turns out, we're omitting water, water quality, and major aqua-exports—a game changer for everyone. The *evidence* can be seen in the decline of the middle class since the 1970s.

$$\bigcirc > \$$$

In 2015, the Federal Reserve stated that its target for *core inflation* was two percent—at the time, the US was well below that level. Economists claimed that prices paid for goods and services then were equal to prices paid in the 1960s. It certainly didn't feel that way, so what did that mean in real terms to us?

Well, what did Legos cost in the 1970s and what do they cost today? What about a candy bar's cost from the days of President Nixon to the years of our current president? How about the change in healthcare costs? These examples would reveal that inflation is working really hard, and on overtime lately, yet we are spending more for less. Has anyone seen how small a Lego set is and how much it costs? The truth is that the prices of goods and services are *adjusted* or *increased* for inflation, but *for the most part* the buying power of our paychecks are not. Our paychecks are not adjusted, nor increased to match those same changes, and our income is constantly losing an uphill race. Meanwhile, the quality of the products we buy decrease in value as inferior materials are *substituted* into the manufacturing process and we are given *less of the same good* for an equal or higher price. We ask ourselves, does the minimum wage increase more often than the price of Skittles? Something doesn't make sense here. Something is missing.

So economists might say "wages are sticky" and prices are not, meaning our wages are noticeably stubborn, while prices we pay always rise—just like they have for candy, healthcare, and Legos. These same economists would argue that wages should increase under conditions of steadily rising inflation. But that's not the case. Life is so much more expensive than in the 1970s. We are experiencing a continuous loss of buying power. The irony behind this dwindling buying power is that we defend it by adjusting prices higher to meet inflation even though it is not allowed in our paycheck.

But what if inflation was skewed to benefit the good name of the US economy and Brand USA? If that were true, the economy would be stealing from us. If we could simply give ourselves a raise by increasing our buying power, is that something we would be interested in? What if increasing our buying power could also benefit those in underdeveloped countries? Would that interest everyone?

The way we do that is by putting water back in our money.

$$\Diamond > \$$$

If all this is true, then where's the *evidence*? In the 1970s, 1980s, and 1990s the CPI robot *included food and energy*. Today, this antiquated robot is known as *headline inflation*. Since the beginning of the twenty-first century, headline inflation is *not* what the Federal Reserve has focused on to determine interest rates—also known as monetary policy. The Fed drives monetary policy with *core inflation* on its dashboard, a CPI which *excludes food and energy*—removing 90 percent of our water usage, or excluding the most water-intensive aqua-exports.

When we think about food and energy, imagine two giant machines which burn water to fuel their engines. Imagine that these two machines demand 90 percent of our water supplies. Think 70 percent for food and 20 percent for energy. The *evidence* shows up between 1971 and 1977, when the CPI rose 47 percent by some estimates. In real terms, this meant that the cost of living (food and energy) was screaming higher. The American dream was getting much more expensive, because regulation was creating more water value in the economy, and benchmarks which measure that change, such as the CPI, were demonstrating this fact *at that time.* Water-intensive food and energy were responding to a new water value brought on by the EPA.

So, the CPI increased 47 percent. What does that mean to us and what does the CPI really do? The Bureau of Labor Statistics described it this way:

> The CPI is often used to adjust consumers' income payments (for example, Social Security) to adjust income eligibility levels for government assistance and to automatically provide cost of living wage adjustments to millions of American workers. As a result of statutory action, the CPI affects the income of millions of Americans. Over 50 million Social Security beneficiaries, and military and Federal Civil Service retirees, have cost of living adjustments tied to the CPI. In addition, eligibility criteria for millions of food stamp recipients, and children who eat lunch at school, are affected by changes in the CPI. Many collective bargaining agreements also tie wage increases to the CPI.

We call cost-of-living adjustments COLAs, and they are based on versions of the CPI. For example, if the CPI were higher, then Social Security payments (COLAs) would rise to offset inflation. *And any budget deficits would become larger.*

How could the US afford to adjust for a 47 percent price change without mounting new debts and deficits? This was the situation in 1978. Policymakers have certainly understood the concerns behind high inflation, especially if they were not prepared to compensate Americans for that change in the prices of goods and services. The US government and the American people care about price. We care about the price inside our bank accounts, the price of goods and services, the price of raising children, and generally the price of our lifestyles. Why is this coming up? Because a realistic price change for the American people seems to be a conflict of interest for the leaders of the United States. If people perceive that inflation is exacerbating the cost of living for Americans, while the data by the CPI robot doesn't communicate a radical change to the public (COLAs), then there is a realistic financial gap in the American dream. And perception can shape reality. This is the beginning of the water gap.

$$\Diamond > \$$$

In 1978, software upgrades to the CPI robot began. The updated robot now calculated "weights from a 1972–73 *survey* of consumer expenditures and the 1970 census," according to the BLS.

In the early 1970s, the US was in a recession, meaning the *surveyed* prices of many goods and services were abnormally cheap or discounted. That's the kicker. The robot formatted 1978 price changes for Americans based on a Black Friday event during a recession. How could Americans expect to afford living expenses based on Black Friday sales prices that occur once every ten years? Their paychecks surely could not. The CPI number gave the economy a much lower number than the prices Americans were paying. The CPI had risen 47 percent in 6 years between 1971 and 1977, yet the CPI revised in 1978, *which was used for the next ten years*, only acknowledged price change from two of those years. By inputting prices from Black Friday, the 1978 revision nearly eliminated or muted two-thirds of reported inflation. These revisions, as measured by the US government, began to drastically invalidate the ripple effects of the EPA acts. Aqua-exports were rapidly rising, yet another story was being told inside our paychecks. There was a disconnect between water value and the CPI that occurred alongside the formation of the EPA.

The US economy had drastically changed. Corporations and their manufacturing plants had left the country, causing economic weakness. And the price change reported to the people, to the Federal Reserve, and to investors took a step backwards in time, to a time before water value's birth in the 1970s.

Simply put, the prices of aqua-exports—the price of everything with water in it—was climbing, and the paycheck values calculated by the CPI robot were printing a number which did not reflect the same change. This was a stealth tax upon Americans. Times were stamped with the economic condition of "stagflation," which occurs when inflation rises, but income, economic growth, and jobs don't—the opposite of what most economists would expect from steadily rising prices. A nightmare formed in the American dream. Water value was being massaged out of paychecks. A little inflation is good for growth, but too much of anything is bad for everyone.

⬦>$

After 1978, more aqua-exports had meaningful tantrums in price. Another oil crisis occurred in 1979. By 1980, US citizens experienced the highest recorded CPI of 13.5 percent, according to the Federal Reserve Bank of Richmond. It was crystal clear, for the first time since World War II, that buying power was *evidently* shrinking. Using history as our proxy, whether or not attempts to mute inflation were *recognizably* successful, aqua-exports rose in price to express new water value. Hence, US dollars could buy fewer and fewer of them. Aqua-exports were responding to regulation and growing water demands from increasing populations. While the EPA was slowly assigning water value to the environment, very few Americans were aware that this was creating water inflation and eventually a water gap in our wallets. Unbeknownst to us all, the CPI robots kept cranking out numbers with less and less water.

⬦>$

The next round of software upgrades for the CPI robot should have reflected the price change from 1974 through 1978, since the

1978 revisions did not capture those weights. But the next revision in 1987 did not account for the inflated numbers from 1974 to 1981. *Once again*, the BLS professors went shopping for data on another Black Friday sales event. The major CPI upgrade instead took weights from the 1982 through 1984 Consumer Expenditure Survey and the 1980 census, according to the BLS. This meant the robots measured price change from 1972 to 1973 and 1982 to 1984.

At the time prices were measured, the US was in *another* recession. The CPI took 1982 to 1984 prices into account for a 1987 lifestyle. This survey of prices more or less omitted any changes from 1974 through 1980—*years the CPI and inflation together were making the highest highs in history*. Nosebleed inflation was caused by the first years that the EPA was fully on the scene, dictating an unrecognized water value. Simply put, *the extreme lows in prices were measured, excluding the extreme highs*.

What does that mean? For example, if the CPI measured the price of oil from 1972 to 1977, the price change was large. If the CPI measured the price of oil from 1972 to 1980, the price change was even larger. But the price change between 1972 and 1974, and then 1982 and 1984, recorded the least change. When the *very important extremes* are muted, finding reliability and value in the data is like sourcing results from the game of "Telephone."

The *evidence* of a 1987 CPI software upgrade using 1982 to 1984 prices is highly suspect, especially since the Fed later referred to this period as "the 1987 inflation scare." The years between 1982 and 1984 offered relatively cheap prices for aqua-exports during the decade, especially water-intensive food and energy. Remember that the US dollar was collapsing in 1980 and the price of aqua-exports, the CPI, and inflation were hitting all-time highs. So Federal Reserve Chairman Paul Volker dramatically hiked interest rates. Increasing interest rates is supposed to curve, or decrease, inflation and vice versa. As we'll learn in the chapters ahead, the United States at that time was experiencing reverberations from the *Latin American Debt Crisis* in the early 1980s.

And these reverberations—felt by both North and South America—were actually linked to Fed coach Volker's interest rate hikes.

Higher interest rates create an incentive to park money rather than spend it. Rising interest rates slow down investors, who choose to freeze and leave liquid dollars in the bank to collect interest rather than invest in risky alternatives such as commodities and the stock market. Thus, aqua-exports were falling in response to short-term interest rate hikes, causing the US dollar to rise. Investors sold commodities and, in essence, bought or demanded the US dollar. During the time price levels were measured by the CPI robot between 1982 and 1984, prices had fallen significantly since their rise in the late 1970s. Yet again in 1987, the CPI conveniently recorded goods and services when price change was discounted and depressed to some of the lowest limits for the decade. After that, the only option American prices had was to rise. This was a well-timed price lock for the government, which has to pay millions of Americans based on COLAs all determined by that CPI.

It's easy to understand how we can be unaware of revisions and software upgrades to the CPI robot. And it now makes sense that the economy surrounding us could digest a discounted version of the CPI for the government's budget and a less expensive COLA, but prices can *still* rise for Americans. As we know, the CPI data recommends what our incomes and salaries should be adjusted toward. So life gets more expensive while our government-stated inflation and wages stay "sticky."

It's like playing the game of Telephone. From the first time a message is spoken to the last relayed communication, the interpretation becomes more inaccurate. Every time BLS professors touched the data, something got lost in translation—that something was water. Downstream consumers and families are traversing this covert-inflationary desert, hoping to achieve the middle-class dream, as if seeking an oasis. Costs are not what they seem because inflation is not what it says.

Water inflation yet again was compounded in 1987 as the change in price of aqua-exports was minimized and muted for the CPI reading.

Thus, a water gap was being further carved out *once more* from the American dream.

◊>$

Marvin Goodfriend, the Federal Reserve's senior vice president and policy advisor from 1993 to 2005, stated a need to restore the Fed's "credibility for low inflation that was lost in the second half of the 1980s." In other words, *either the Fed's monetary policy or something the Fed was basing interest rates on (the CPI) was not credible.*

One pillar of intent by the Federal Reserve Act as laid out by Congress is to maintain proper levels of inflation and create price stability. The other tenets of the Fed are to moderate interest rates over longer periods and maximize employment. Boiled down, inflation, interest rates, and jobs are all highly connected to price change, a.k.a. the CPI. And price is what we should march to. If we follow the money, we follow price, and if we track price change, we'll understand why the American dream feels so expensive.

In hindsight, after two decades, it appears that the professors at the BLS used a "curve" that locked prices for the CPI in the 1970s and 1980s to Black Friday recessions, as some of the extreme highs in inflation were omitted. As noted by Goodfriend, there was tension about the effectiveness of the Fed's *interest rate response* to reported inflation. For example, if the CPI was actually muted, then interest rates were also muted and the Fed's *response* was insufficient to tame *perceived* inflation. If the CPI does not include the full impact of inflation on price change because prices were recorded during depressed Black Friday sales, then the CPI did not translate some water value, creating a water gap. That gap explains the difference between government-stated inflation and publicly perceived inflation. Simply put, water inflation created the reality of higher prices. If we remove water value from economic readings for price change, we remove it from our money, thus our interest rates are also not responding to real inflation.

Therefore, our interest rates were misled and too low. Going into the twenty-first century, this affected our housing market.

$$\lozenge>\$$$

History simply repeats itself, doesn't it? The next round of CPI revisions should have been timed to another Black Friday sales event, right?

Well, that's exactly what happened.

Once more, the CPI robot received software upgrades in 1998 and the BLS took price changes from a 1993 to 1995 consumer survey, according to the BLS. It was a period marked by the words "disinflation" and "recession." For example, in 1994, oil was less than $16 a barrel. But prices do not remain that low forever.

In fact, Marvin Goodfriend felt it necessary to explain this in a quarterly report, stating, "The Fed's credibility for low inflation had been compromised twelve times from 1987 to 2001."

After 1998, the CPI robot told a story about price change between 1993 and 1995, but not the prices Americans were paying at the time. Washington may have never recognized a new water value due to the impact of the EPA's policies and laws, but political heads certainly noticed the interpretation of price change on the CPI. With all the robots' efforts, the CPI was *still* too expensive for programs like Social Security. Simply put, not enough water was removed from price change. Pressure was building to have the CPI robot upgraded to a new operating system or replaced by a new model altogether.

A fundamental shift by the name of the "Boskin Commission" occurred in the way water costs were translated into inflation. A significant report presented before the 1998 revisions now exposed the CPI to new upgrades through a fresh set of tools. The outcome exacerbated hidden water inflation.

This new type of approach could mean that any future Federal Reserve decisions on interest rates based on CPI data wouldn't make a strong enough connection to water as the economic driver to life. That would falsely deflate price change, underrepresent cost-of-living adjustments, remove more water value, and increase the perceived gap in inflation. Clearly, the CPI did not adjust for a water inflation, but Americans most certainly paid for it.

◊>$

Appointed by the US Senate, the Boskin Commission issued a report in 1996 on how best to measure inflation. It was formally known as The Advisory Commission to Study the Consumer Price Index. The results from the Boskin Commission Report widened the water gap with techniques that created artificially lowered inflation, which spurred artificially lower interest rates. Americans were forced to pay for the higher cost of aqua-exports while the US government downwardly adjusted inflation and pointed to the benefits of lower interest rates. Some responsibility could fall on the findings of the Boskin Commission and its effect on the housing bubble and collapse in home prices in the twenty-first century.

What fundamental changes to the CPI caused this? What new tools did the Commission use to justify this reduction in the cost of living? How was water removed from our wallets? There were three tools: *quality change*, *substitution*, and *hedonic adjustments*. After this report, there were lots of questions regarding them all.

Many well-qualified individuals felt that quality change was "subjective" and poorly researched. For example, quality change emphasized the importance of innovations in computers, mobile devices, and the picture quality of our television sets, rather than the quality of the food we buy daily.

In the aftermath, American-grown food could be *substituted* for lesser-quality foods from other countries with very lax environmental regulations. And the way the American people were compensated was through the same TV prices but with better-quality pictures and thinner technologies. *The Boskin Commission placed more emphasis on what was on sale during Black Fridays than what we ate at Thanksgiving.* It was about TVs and not turkeys. So while a TV is an aqua-export and requires water in production, nothing is more water-intensive than food, especially meat. And nothing affects us more.

Did the American people have a choice on quality change or substitution? Would we rather have our paychecks adjust higher for more expensive food *or* enjoy the same prices for better TVs we purchase years and years apart?

This quality change led to substitution. And substitution later allowed for hedonic adjustments. Barry Ritholtz, founder and CIO of Ritholtz Wealth Management, newspaper columnist, author, and television commentator, had this to say on the matter of substitution and the Boskin Commission:

> Substitution is a nonsensical approach that adjusts inflation for consumer behavior. When steak prices rise, consumers "substitute" cheaper proteins such as hamburger or chicken. Thus, Boskin states, the consumer is spending no more than they previously were, and is not suffering inflation. The reality is that consumers have been priced out of steak due to price increases. Oh, and somehow, the decrease in quality does not get hedonically adjusted when it raises inflation.

As policymakers created new means to explain the times, was this an indication that wages couldn't rise? Was this a warning that the American dream was losing its steam? The Boskin Commission could have been confessing that substitution was a more effective and logical means to adjust for our higher cost of living. Can we assume that when water-intensive steak rose in price, the Commission indirectly assumed that people would not work harder to earn steak, but settle or substitute for hamburgers or cheaper chicken? Did this replace a raise? Was substitution a solution for our paychecks, which were not keeping up with our everyday costs? Steak is, after all, one of the most water intense aqua-exports, even when compared to chicken. Could we simply not afford as much water?

In visiting the Commission's hedonic adjustments, Ritholtz followed up:

> Hedonic adjustments are addressing the improvement in quality as a form of deflation. For example, the price of a new car in the US had risen from $6,847 in 1979 to $27,940 in 2004. Using hedonic adjustments, the government calculated the price of a new car had risen from $6,847 in 1979 to $11,708 in 2004.

The innovation in safety and gas mileage allowed the BLS professor to downwardly adjust the actual price of a car to an unrealistic number in the CPI data. But the BLS professor didn't pay $11,708 for a car. This means hedonic adjustments allowed for a "subjective" or "opinionated" deflation in price measurements—a false deflation. In reality, we all paid on average a rounded $28,000 for a new car in 2004, not $12,000. Not only that, but our buying power was substantially lower in 2004, because the US dollar was much stronger in 1979— roughly 20 percent stronger.

So, Ritholtz said, policymakers adjusted CPI through substitution, supported by hedonic regression. Okay, but we still don't get it, and that's exactly where policymakers expect us to give up. For the purposes of Ritholtz's comments, hedonic adjustments are deflating prices on the grounds that improvements in quality and time savings cancel out inflation. Is that true?

As products are introduced, quality of life could improve or deteriorate depending on perspective. For example, a professor at the BLS may drive a very safe Lexus hybrid; however, they could also be diagnosed with stage four cancer. The Lexus did not improve the quality of life; it just assured driver safety and fuel efficiency. Meanwhile, life just got very pricey for this BLS professor, and a car had nothing to do with it. The professor's paycheck was not adjusted for cancer.

If the quality of inputs for our lives—such as food quality—takes less importance in the way inflation is measured, so does our buying power for what really counts. No matter how fast your Internet is, if your food intake is associated with chemicals linked to cancer, then your quality of life goes down. Disease kills time, even though timesaving products, like miniature smartphone-computers, seem to have a robust importance.

The chart herein highlights the extreme discrepancy between the upgraded CPI with all its new bells and whistles and the reaction of commodities that are so water intensive. Since the dawn of the twenty-first century, aqua-exports look like they felt disregarded by all the CPI updates and reacted in a passive-aggressive move higher. The theme following Boskin is that aqua-export prices are rising and CPI and interest rates are falling.

Chart courtesy of StockCharts.com

What do we notice in the commodity chart about the years following the use of new Boskin tools after 1998? Yes, prices climbed for commodities after the new methodologies and software upgrades were active. Aqua-exports were obviously speaking for something not included in the CPI—something that got lost in translation. A majority of the water value thereafter was omitted in the way inflation was measured, and instead was digested in the price of aqua-exports. The most staggering conclusion of the Boskin Commission was that inflation was overstated, meaning the US government was overpaying Americans and their paychecks by 1.1 percent. So we took a pay cut.

$$\mathbin{\Diamond}>\$$$

Because interest rates and inflation also determine a currency's value, every new methodology and software upgrade removed more and more water from the value of the US dollar, and so it fell. Water inflation caused commodity prices to explode to all-time highs in 2008. Yet the media failed to recognize the growing water gap or resulting higher water value affecting commodities. This is a pattern that persists today, eroding our real wages. There is no denying that yearly expenses are rising faster than our stubborn wages, without a new, costly diploma and specific career change. Or that our wages are not rising as fast as water inflation, creating a water gap in our wallets, for which we all pay.

Generally, we lower interest rates to spur economic growth. A lower interest rate inspires economic expansion because it is cheaper to borrow and spend money. During economic growth, traditional economists would expect inflation to rise. They would anticipate that the Fed would increase interest rates to cool off rising inflation. Intuitively, higher interest rates slow the economy and lower inflation, and in some cases cause deflation. Overall, interest rates and inflation seesaw. We know from history that the two are highly correlated. First inflation goes up, and then so do interest rates. Then inflation falls and interest rates follow. *However, when water inflation was not included in consumer readings for the CPI, it created falsely-deflated interest rates in the United States, generating a water gap in monetary policy.* Interest rates rise to curb inflation, but if water value is missing from inflation, then inflation readings remain too low and thus interest rates remain too low. Therefore, the Fed has a gap in monetary policy.

The CPI upgrades and Boskin Report made it possible for interest rates and inflation to *decouple* going into the twenty-first century. Therefore, a portion of water value officially had no direct power over the price of goods and services, and aqua-exports had no other choice but to express water inflation in their price. And so all commodities soared. The CPI readings omitted so much water value, and this gave the Fed space to lower interest rates to spur growth, albeit a false growth. The type of water inflation that hit the American people should have been met with sincere interest rate adjustments higher, but water went missing from the CPI, and therefore water went missing from money.

CHAPTER 6

Apples to Apples ⬤

WE WORK FOR A living in order to afford choice. But are our options, as measured by government, the same choices we face on a daily basis?

We have television sets today relatively the same price as those of the 1980s, but better, smarter, and thinner. Yet we don't buy TVs every year. Cell phones are now computers, and they all offer video conferencing. But we don't buy cell phones every month. Yes, technology has improved our convenience and time, and brought us closer together globally . . .

But what about cancer institutes? The facilities are becoming more advanced and more expensive simultaneously. Our healthcare costs are also rising. From 2013 to 2015, healthcare costs rose 25 percent, according to the Oxford Club. *But core government price readings said inflation numbers were well below two percent.* The same Club shared food inflation per annum circa 2014:

REAL INFLATION

Beef and Veal	+22.5%
Ground Beef	+21%
Steaks	+14.9%
Pork	+7.4%
Ham	+11.5%
Whole Chicken	+6.1%
Fresh Fish	+3.5%
Eggs	+8.2%
Cheese	+7.8%
Fresh Vegetables	+4.3%
Lettuce	+12.2%
Tomatoes	+9.6%
Coffee	+6.7%
Butter	+19.5%

Source: Wealth Report (Oxford Club)

Essentially, we're being told that our cost of living is flat, but in reality, it is going way up on the stuff that matters, even as technology gives us more choices.

So let's consider food quality. Food expenditures that are affordable and measured in the basket of the CPI have become the victim of globalization. Animals spend their lifetimes lying in their own feces, packed together in dark cages with sharp wire, injected with growth hormones, and fed with the same fertilized compounds as explosives. What we eat is washed in herbicides, grown with genetically modified organisms, showered in pesticides and preservatives, and fed with contaminated and polluted water supplies with heavy metals to boot. All these practices are linked to cancer. The food of 1913 did not taste like the food of today, and it sure did not include the same chemical feedstocks and water quality. Today, if a food doesn't have high levels of inorganic chemicals within its growing process, it is certified and labeled.

This certified food costs a lot more money than what the CPI suggests—thanks to *substitution*. Sure, we need to feed the growing billions against drought conditions, but the people in developed economies are paying many different prices for food and receiving different qualities. Those who pay the least for food now could pay the most regarding healthcare costs later—or not pay their medical bills at all. Those who pay the most for higher-quality food may pay the least for healthcare expenses in the end. But this is obviously subjective. The jury is still out.

To the point, the CPI is not measuring certified, labeled, and more costly quality food in the way tax brackets are assigned for Americans' income. The CPI is treating everyone like they are in the lower tax bracket—as if everyone shops for food at discount retailers like Walmart. But that's not the food everyone is paying for and eating.

Today, there are tiers of produce—quality versus quantity, and "steak versus chicken." You pay for the type of food you want and the protection from cancerous products you don't. Those who are aware do not buy non-certified produce at discount retailers like Walmart. However, the CPI accounts for much of those foods sold at Walmart— roughly 25 percent comes from the big box, a.k.a. the "Walmart Effect." Walmart provides cheap prices by leveraging economies of scale and *least-cost rules*. The same economic leverages were being applied when larger farms consolidated smaller farms beginning in the 1930s. Small farms gave way to large, industrialized farms as chemical applications in farming techniques increased output per acre. Many would generalize some of these techniques today as feeding-the-world strategies, but others would call them shortcut policies from the military-industrial complex. While farming techniques may change over the decades, their footprint certainly does not leave the United States.

Walmart benefited by importing from countries with the least environmental regulation—those nations that were exploiting their environment and these farming techniques, like Guatemala or Honduras for bananas. Chile and China were also forerunners in least-cost rules, or profit maximization. American food importers realized that if a foreign country had no environmental authority, farmers could use more industrial fertilizers, pesticides, and preservatives to produce exceptional harvests. Today, some countries have completely different standards than US farmers, who are forced to comply with environmental and health regulations. A certified apple from a US orchard, for example, must stand above stringent EPA mandates and meet US Food and Drug Administration standards, avoid genetically modified organisms (GMO), and adhere to organic requirements. All these standards mean that an apple grown with US heart and precision is much more expensive than the one imported from Chile or China. Yet which one is used for measuring the core CPI?

Cheap, least-cost apples from abroad are used as the benchmark for consumer prices in the United States, and undervalue the cost of domestic apples we actually buy. As a result of the many decades of CPI upgrades, quality change, substitution, and hedonic adjustments, we have a dilemma. Our paychecks and potential raises are secretly based more on applesauce from China than on the price of apples we choose to buy from the grocery store or local farmers market.

$$\bigcirc > \$$$

Water is in everything, and connects us all, one way or another. As we remember the Vietnamese children suffering from Agent Orange, we realize that so many of our foods cannot escape the history of their groundwater. So, would we agree that how we measure inflation has *not* been adjusted for water quality? If a car can hedonically adjust inflation downward from $28,000 to $12,000, why can't a more invasive product like water?

The CPI does not hedonically inflate the CPI for more contaminated water supplies the same way it deflates for quality change in cars (safety and mileage). Furthermore, our bodies cannot forget the unrecognized leaks created by contaminated groundwater in our food, or the long-term healthcare costs. These ULs remain in our bodies' DNA for years, morphing into unconscious cancer cells. Yet without justice for the public, a paramount economic benchmark such as the CPI, interpreted by the US Federal Reserve and US Government—which adjusts American incomes and Social Security payments (COLAs)—has *never calculated for water quality* or included a water quality index (neither does the PCE robot that we'll discuss shortly). Unfortunately, we use the core, but if headline inflation—the CPI that includes food and energy—were used, it would still *not express* water value, because even this version of the CPI excludes a water quality index. Hence, we are *forced to live with unrecognized water inflation because we don't recognize water quality in our wallets.*

When we measure the change in the cost of food for present-day prices paid at the supermarket, it's a false evaluation. The price for the contaminated foods we eat is paid in the future, in terms of healthcare costs and sickness. Our history of consumption and our parents' history of consuming food from the 1950s all the way up to the twenty-first century cannot be forgotten.

The same is true with water. Contaminants in groundwater must be measured in terms of their downstream impacts on our health. Consumers need to know about the real value and costs of water we consume in our products. Exposure to arsenic, for example, can cause gastrointestinal problems and skin discoloration or lesions. Exposure over time, which the World Health Organization says could be five to twenty years, could increase the risk of various cancers and high blood pressure.

We must remember that we are what we eat. In the US, the EPA only gives the groundwater that supplies public drinking sources "optional" or "periodic" surveys for "fecal contamination," a process merely implemented in phases based on the "Ground Water Rule." According to a twenty-year US government assessment of groundwater quality by

NAWQA, we don't know whether roughly 100 constituents measured in the drinking water have concentrations which are potential health risks to humans. *One out of every three Americans drinks this groundwater.* On a national average, one of every five samples collected from that drinking water "contained at least one contaminant at a concentration that exceeded . . . the human health benchmark." Likewise, the water supplies used to grow our food are also behind in terms of actual regulation for water quality. If the water we are drinking is a mystery, then so is our food.

◊>$

You may already know that big-box stores like Walmart have tremendous international buying power. This has its pros and cons. Walmart can push all sellers' prices destined for its shelves to the lowest levels. Furthermore, its reach is global and relies on low-cost factors of production. This has a tremendous impact on the prices of aqua-exports (all goods and services). Ten years after the 1996 Boskin Report, Robert J. Gordon produced a paper for the National Bureau of Economic Research. In it, he broke down the aftermath of the Boskin Commission and within his research, he highlights the "Walmart Effect."

If 25 percent of food expenditures come from the "Walmart Effect," that could mean the CPI expects consumers to match their expenses with "Walmart" diets. Additionally, big-box stores do not communicate the appropriate water inflation from markets that have the least environmental regulation such as China and, historically, Chile. What will happen to the prices on Walmart's shelves when all countries institute global environmental regulations that match the EPA's? Would everyone experience major water inflation?

And what about the potential substitution of apples in the CPI for Walmart applesauce? What if we were told that six four-ounce cups of applesauce from Chile or Chinese concentrate is how our apples are priced to determine inflation? At the time of this writing, a six-pack of applesauce cost $1.68 and may represent a range of twelve to twenty-four apples. In some stores today, one apple could cost that much or more. Importing massive amounts of "cheap" food comes with strings attached and a long-term price. Illness. Cheap sources of food keep prices artificially suppressed because we finance the real price, long term. Being sick escalates healthcare costs, lowers work productivity, lowers potential earnings, and threatens our retirement dollars. That's the real price. "Live longer and die slower" is a pattern that has been emerging for decades.

Since the 1970s, more and more food has come from outside the United States—specifically, seasonal fruit from nations such as Chile. So we must factor that change into how we measure inflation and how the environmental practices of those countries become part of our consumption, especially if our CPI accounts for the Walmart Effect.

If the water quality found in an aqua-export such as an apple is imported by a major corporation like Walmart, part of the water inflation will forever be absorbed: first, in the devastating environmental policies upon that emerging market (Chile or China) and people of that nation; then, in the transfer of that commodity's chemicals in the fruit we eat; and finally, in the trade policy.

What do we mean by trade policy? Krugman and Obstfeld explored this relationship with the law of price and commodities. In discussing universal economic agreements, these brilliant men highlighted that the "law of one price" does not hold true for purchasing power parity. They explained how Americans might have paid less for apples or grapes than a country like Chile, which suffered direct and immediate harm in growing them.

"The law of one price states that in competitive markets free of transportation costs and official barriers to trade (such as tariffs), identical goods sold in different countries must sell for the same price when their prices are expressed in terms of the same currency," they wrote. "Prices of identical commodity baskets, when converted to a single currency, differ substantially across different countries . . . Manufactured goods that seem to be very similar to each other have sold at widely different prices in various markets since the early 1970s."

Perhaps what Krugman and Obstfeld did not consider was an unrecognized water value in each of those markets. For example, the US's unrecognized water value was much higher than Chile's unrecognized water value post-1970 after the EPA's impact. That means that not only was trade equality off in terms of currency, but the use of water was not evaluated then, making any discounts for fruit even more extreme to the US, which was clearly valuing water differently.

Applying this economic theory to Walmart's kind of international purchasing power means that a Chilean citizen purchasing a Chilean apple could experience price discrimination, as opposed to a big-box customer in the US buying applesauce priced below market value. Might those be the same apples in the CPI?

This again bids down and distorts the global water value picture and instigates water inflation everywhere. We are underpaying upfront and overpaying later. Chile is simply overpaying, period, and as we'll learn, its economy proves that.

$$\triangle > \$$$

We know water inflation is rising with the opportunity costs of other industries rationing supply in the United States. Yet we can underpay for an apple imported from Chile and still experience its water inflation in healthcare dollars. In sum, falsely deflated goods will become realistically priced once other nations end the sale of cheap water with new environmental regulation. What will happen to the middle class in developed countries like the US, which have become very accustomed to falsely-deflated goods and services coming from these countries?

We'll experience water inflation, as global water supplies become regulated and ultimately change the price of everyone's goods and services. We'll find ourselves living longer in a more expensive world. But will that longer life be a better life?

Over time, the CPI has masked water's direct connection to our economy and thus removed its upstream value from its downstream output. It's safe to say that the CPI is very complicated in how it's reformulated and revised through each succession of years. It may now feel more natural to agree that as water is used more, its impact on price has also been limited and its voice for quality overlooked in terms of real inflation. Something is certainly lost each time the professors at the BLS touch the data. When the benchmarks that gauge inflation become overly complex or outright replaced, will our middle class progress or regress through the lenses of water? Water is in everything, and connects us all, one way or another.

CHAPTER 7

Water Is Volatility ⬤

IN THE YEAR 2011, THE prices of aqua-exports (commodities) were rocketing to the stratosphere, and CPI was underperforming according to its critics. "Accommodative" or expansionary money printing by the Fed coaches should have explained why inflation would rise—but didn't—and droughts around the world were ramping up to extremes simultaneously. It didn't make sense. Something was malfunctioning. Gold hit all-time new highs at the end of 2011, and silver matched its 1980 peak of $49 per ounce in the spring of that same year. Meanwhile, water-intensive corn marched through all historical price ceilings in 2012, and wheat followed by making fresh peaks as well. Oddly, in the same year, the Federal Reserve swapped out its dashboard for inflation with a new set of professors and a new robot. Everything about this change was convenient.

What about history and consistency? Nothing was more consistent than the time prices were collected for the CPI, including the new inflation dashboard in 2012. Professors over at the Bureau of Economic Analysis (BEA) ushered in the PCE robot (Personal Consumption Expenditures). In 2012, when the CPI became too much of a victim of investment analysts and their attack on how the CPI was calculated, the Federal Reserve simply switched its attention from the *core* CPI (excluding food and energy) to the *core* PCE (excluding food and energy). But here's the kicker: *the price change Fed coaches used to measure inflation through the core PCE was benchmarked to 2009, the worst price collapse since the Great Depression, and yet again, price change was locked to another Black Friday sales event.*

Was the robot switch indicative of too much error, touches, and upgrades? Cutting attachments to the CPI allowed policymakers to start fresh from the housing bubble and financial crisis in 2008 and reduce any exposure to policies and relationships sown by the CPI in the past.

In 2012, PCE became the new inflationary number for "Fed speak," yet the damage and gaps were already complete from the years of the CPI as ringleader. Ironically, this same year was building up to the worst drought since 1956. With even more irony, this benchmark, the PCE, was indexed to prices in 2009 after the biggest price decline of all goods and services since the 1929 stock market crash. In 2009, aqua-exports were destroyed in value. Yet this was the year the PCE robot was locked to memory. Another Black Friday sales event for price changes in history. Proof that history simply repeats itself.

Even more important is that the core PCE robot today reflects significantly less inflation than the core CPI when compared—an outcome that benefits the US government but doesn't compensate the American people. Yes, both measures have something in common. Both are "core" indexes that exclude food and energy. It does not matter what the professors at the BEA or BLS offer as benchmarks; our economic wherewithal depends on what the Fed chooses to see. And the Federal Reserve prefers the "core." Remember *AIR? Agriculture* is that giant machine that needs roughly 70 percent of the world's water supply, and *industry* is that other giant, which requires at least 20 percent of the world's water. We're talking 90 percent of the global supply. And these machines' demand for water is actually growing too fast. This water demand from food and energy will exceed 100 percent in the early decades of the twenty-first century, and then what's going to be left for the people? Us *residents* are going to need some water, too.

Because core CPI and core PCE simply exclude food and energy, we can most certainly make the simple argument that the Fed will not include 90 percent of water in terms of how it measures inflation. Why would the Federal Reserve do that? Because the Fed wants to remove the "noise," a.k.a. volatility. Unfortunately, that volatility ends up being water.

The chart below indicates what the "core" inflation index looks like without water-intensive food and energy. The "headline" inflation visualizes how the government once adjusted our wallets, when we included food and energy price changes. The core CPI is minus 90 percent water; the headline inflation is *100 percent water*. We know the Fed prefers the core.

INFLATION DIFFERENCES
a.k.a. Price Change

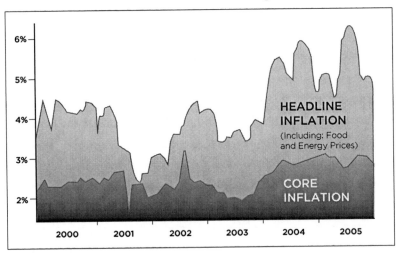

If the Fed only took its cue from the core CPI reading, which measures inflation and inspires interest rate adjustments, it would exclude water value from our diet. The most volatile items are food and energy, which are also the most water-intensive items. Just like blizzards, droughts, and floods, water is violent on the economy. The twenty-first-century culture of the Federal Reserve is to remove volatility from all economic inflation readings. *Water is volatility.* Whether the Federal Reserve is aware or not, the Fed's decisions seem very concerned with water's force as it relates to the efficacy of its policy. So the Fed removes water and volatility from its readings. And why is that?

Michael Pento, senior economist at Euro Pacific Capital, explained through what he called "core incompetency."

> For years the Federal Reserve has told us that in order to detect inflation in the economy, it is important to separate "signal from noise" by focusing on "core" inflation statistics, which exclude changes in food and energy prices. Because food and energy figure so prominently into consumer spending, this maneuver is not without controversy. But the Fed counters the criticism by pointing to the apparent volatility of the broader "headline" inflation figure, which includes food and energy. The Fed tells us that the danger lies in making a monetary policy mistake based on unreliable statistics. Being more stable (they tell us), the core is their preferred guide. Sounds reasonable—but it isn't.

> If it were truly just a question of volatility, the Fed may have a point. But for headline inflation to be considered truly volatile, it must be evenly volatile both above and below the core rate of inflation over time. If such were the case, throwing out the high and the low could be a good idea. However, we have found that for more than a decade, headline inflation has been consistently higher than core inflation. Once you understand this, it becomes much more plausible to argue that the Fed excludes food

and energy not because those prices are volatile, but because they are rising.

If you talk about the grand sweep of Fed policy, it's fairly easy to fix the onset of our current monetary period with the onset of the dot-com recession of 2000. To prevent the economy from going further into recession at that time, the Fed began cutting interest rates farther and faster than at any other time in our history. During the ensuing 11 years, interest rates have been held consistently below the rate of inflation. Even when the economy was seemingly robust in the mid years of the last decade, monetary policy was widely considered accommodative.

Over that time, annual headline Consumer Price Index (CPI) data has been higher than the Core CPI 9 out of 11 years, or 81 percent of the time. Looking at the data another way, over that time frame, the US dollar has lost 20 percent of its purchasing power if depreciated year by year using core inflation, and 24 percent if depreciated annually with headline inflation. The same pattern held during the inflationary period between 1977 thru 1980, when the Fed's massive money printing sent the headline inflation rate well above the core reading. The empirical evidence is abundantly clear. When the Fed is debasing the dollar, headline inflation rises faster than core. The reason for this is clear. Food and energy prices are closely exposed to commodity prices.

On the surface, the Fed discloses that volatility is not welcome. Below the surface, we know that's because volatility represents water's force on the economy. Pento claimed, "The Fed excludes food and energy not because those prices are volatile, but because they are rising." Why is that? Because unrecognized water inflation is rising.

Boiled down, food and energy represent 90 percent of our water diet and they are most exposed to water inflation. The Fed excludes water inflation. And so only we suffer from it.

If there is no recognized water inflation, how could any real inflation ever be publicly recognized or factored in to rebalance our wallets?

We can go further and say that not recognizing water inflation allows the United States to do many things besides just punishing its own currency. According to Pento's findings, interest rates are below the rate of inflation. Thus, interest rates have been allowed to decouple from any realistic view of the economy. The mirage of lower interest rates allows for loans that should never be funded, as was the case with the US housing bubble in the first decade of the twenty-first century—later leading to a collapse in home prices.

Pento references the "recession of 2000" and the timing of interest rates decoupling from inflation because the recession followed the 1996 Boskin Commission results and the profound software upgrades to the CPI in 1998, 1987, and 1978.

We shouldn't lie down and give policymakers so much latitude with inflation or permission to exclude water inflation, because those numbers decide interest rates and create bubbles. When bubbles pop, they hurt the residents the most. Decisions by Fed coaches must embrace the consequence of water supplies and connection to our water quality.

More importantly, if inflation is supposedly low, then economists can say growth is low or normal, and the Fed can justify printing money in the act of self-deprecating currency wars. Whether there's "actually" growth in the economy depends on measured inflation.

Is it possible to experience inflation and not have a growing economy? Yes—1 percent GDP and a 2 percent inflation is negative, a.k.a. in the red. That's called "taking a loss," and it's *not very American*, is it? It's a recession or depression. And that's exactly why it's "un-American" to tell the whole truth about inflation numbers. If massaged, numbers can tell a different story, rather than the reality. It's simply a headline for our eyes. "Figures don't lie, but politicians do figure." We get upset when robbers break into our home and steal our family's artwork, jewelry, and valuables. The irony is that an *unrecognized* recession that only individual households absorb is like a robber never leaving our home. And we experience a constant recession if the robots are missing water.

So, the Fed excludes water inflation and the volatility of water-intensive aqua-exports and therefore *reduces* the variability in our economy. According to the Richmond Fed, "Some empirical evidence indeed suggests that policies that reduce inflation variability are likely to promote economic growth, independent of the inflation."

That means the Federal Reserve has preferences that it knows promote robbery in our homes.

And if inflation is artificially low, then no matter how the economy or individual households are actually performing, interest rates can be falsely adjusted and lowered to stoke growth with cheap money. Whether that growth is real or misrepresented and whether new growth is necessary are beside the point. Perception is a priority to the US economy.

If growth is perceived in the economy, it's because we believe we are growing, and it helps us believe by seeing those numbers printed by professors at the BEA and BLS. Interest rates at an all-time low force us to believe that the only place the economy can go is higher. But what many people do not realize is that interest rates can be negative, and they have been in Europe, meaning the people of Europe pay to hold onto their paychecks. If that's not stealing, I don't know what is. If the economy has zero growth and there is a real inflation, that would be the definition of a long-term depression. If the economy has zero growth and there is real inflation and interest rates go negative, that's *really* stealing from *us*.

So, by muting water inflation on the grounds of volatility, we are promoting false growth, with false interest rates, a recession in our home despite the headlines, and robbery to ourselves.

Meanwhile, pretending the economy is growing makes us feel safe in spending money. Sustained economic recoveries only occur when the American consumer spends money. Whether their buying power is stronger or weaker in the long run is irrelevant to the government.

A CPI reading without water inflation is falsely too low. A CPI reading which omits a water gap caused by water inflation is falsely too low. And furthermore, a CPI reading is falsely too low if it does not account for comprehensive water value, including a water quality index.

Oft-mentioned, a CPI or PCE reading without recognized water inflation allows the Federal Reserve to lower interest rates to promote a false growth. The lower the CPI or PCE, the more aggressive in lowering interest rates and money printing the Federal Reserve can be. More money printing through the mirage of lower interest rates allows for such rash loans as those that fueled the US housing bubble.

The American public is screaming for higher wages to meet the higher cost of living, but the dashboards for inflation are saying there is no reason for wages to rise. Economic tenets profess that if there's not a steadily increasing rate of inflation, then a raise in actual wages cannot be justified. *No inflation, no getting a raise.*

There is a mystery we need to solve. If we cannot find inflation through the lenses of the Federal Reserve, we cannot receive a raise. But if we know there is substantial water inflation in our economy, then we know there is a water gap affecting our lifestyles. If there is a water gap, then the economies around the world are missing meaningful water value from all economic readings. Gaps need to be filled, one way or another. How is that going to play out? In the end, we'll have two choices: one choice perpetuates robbery and the other a thriving economy.

$$\Diamond > \$$$

It seems interest rates have decoupled from the real economy since the onset of the twenty-first century, and it's clearer how policies remove necessary water from economic underpinnings. If the CPI and

PCE robot have a gap in their data, then we should agree that interest rates have decoupled from inflation because the economy is missing the most important connection—water. If economic and financial gurus are constantly questioning why rates are too low, perhaps it is because we need to step outside the box and put water back into our money.

The US Government offers so many readings on inflation that it does not matter which one the Federal Reserve uses for precedence. The Fed, the BEA, or the BLS will always point us to another index, with a more complex formula for gauging inflation. Moreover, no one can beat the Federal Reserve at a game in which it chooses the numbers. The former US Republican of Texas serving the House of Representatives, Dr. Ron Paul, can testify to that. Paul called the housing collapse well before the bubble was ever close to popping. He rose to political prominence, challenging the Federal Reserve daily like a martyr for the people.

All we are doing here as we learn about water value and water inflation is telling the Fed, "We understand true economic operations, even if you don't." It's time to itemize our concerns with water value and then offer one another the choice, using water as the tool to rebalance our wallets.

$$\lozenge > \$$$

The US is grinding against a dire GDP, saturated by major government entitlements and subsidies, withering in a drought-stricken climate, constrained by shrinking aquifers and slowing rivers, and only met by a fickle EPA. All aqua-export values will throw a tantrum, and digital food lines and digital food stamps will increase, creating more dependency on the government. Nothing will offset this fiscal and monetary debt globally unless we let go of control. Water must rise—*not* privately and *not* quietly, but with everyone watching and expecting their wages to rise, too. The message and practice must be specific.

If we allow one another to believe that water is a commodity and *not* a national security requirement, then we are allowing *technocrats* to make way for water's publicly accepted commodification and ultimate privatization—an outcome that would lead to financial terror on an individual household level, like robbers never leaving our home. At the same time I finished my first round of theories of waconomics in 2006, the message at the end of the film *The Big Short* was born. For our families to thrive, we need to ignore the message of Dr. Michael Burry in the context to which the movie delivered. We need *not* acknowledge Wall Street's value of water. Those who will profit from that message want us to talk about it. They benefit as we discuss it more. These investors want us to believe it's a commodity—so as water ownership quietly increases and water is totally recognized for its value, the big squeeze will continually be on the people. The same way there were profits in the housing market for a big short, that opportunity hurt those who were unaware.

There is a major difference from Wall Street's water value and the US government's water value. Join me as we turn off this movie in our heads and focus on the goal. The goal: do not commodify or fully privatize water; keep it with the city as our *public duty*—just increase the price based on some calculations and data. Place a moratorium on privatization and make every city municipality a *national treasure*. We already sold our roads and our bridges; now water is on the chopping block.

Ask our local municipality to join us—to eliminate surprise by appreciating water's rise. Until a time when our food costs and healthcare expenses are linked to water quality and water supplies, prices slowly rise, life becomes more expensive without matching incomes, and our standard of living outpaces our real wages and declines. A wise individual once told me that he was a Democrat on the local level, a Republican on the state level, and an Independent on the federal level. There are no political parties here, only price. However, if we must apply Democratic or Republican principles to get to true independence, so be it.

CHAPTER 8

Global Conquest, Corporate Commies 💧

YOU CAN SEE THE geopolitics of water play out all over the globe. Europe has been an especially interesting example. Countries in Western Europe prospered because of an abundance of water and aqua-exports for living. When those resources grew scarcer, Europeans did what they always have—they sought them elsewhere. Such was the case in South America.

The southern continent contains 31 percent of the world's freshwater and offers two-thirds of the earth's oxygen via its rainforest. Brazil hosts the Amazon River, a majority contributor to this vast water supply—equal to one-fifth of the world's freshwater. In recent years, the Brazilian economy and currency have grown more powerful by the day because of relative water abundance. But it was a prisoner to European imperialism for hundreds of years. Hence the extraction of water-intensive goods through the agricultural export model, as practiced for centuries.

The Europeans stripped countries in South America of their well-endowed lands, just as Spanish conquistadors genocidally removed somewhere between thirty to eighty million indigenous people from

South America. Even now, as we compare the lives taken in the Holocaust, we realize just how important South America must have been for resources, that so many people died for the cause—at the time, that number represented upwards of 16 percent of the world's people. Fast forward to the twentieth century, and this conquest left the most resource-abundant South American nations open to tremendous currency flux connected to European and US tantrums. Following World War I, whenever resistance to the European or US agenda surfaced, countries who opposed their export supply requests experienced forms of hyperinflation. When countries such as Brazil revolted against the export-led trade rule of hegemonic might, their currency was brutally punished. This robbed the Brazilian people of their purchasing power, suffocating the domestic economy. The latter part of the twentieth century in South America is notable for the ruling powers' achievements in decoupling aqua-export values and a country's respective currency. Lessons learned: resist exports for developed nations, and foreign governments will treat your nation as though it were dirt-poor, with no water and no resources.

In the end, the citizens of the Southern Cone—Argentina, Brazil, and Chile (ABC powers)—were not the main beneficiaries of the European export-growth model. The lion's share of aqua-exports fed the growth in Europe and North America before and after World War II. Economic policies were customized for developed nations and their corporate heirs. The revenue from aqua-export operations was repatriated to advanced nations.

The value of water used for production paid dividends to foreigners and some corrupt political figures within South America. Yet when the South American people wanted to deny Europe and the US access to the export model, the first consequence was extreme inflation. Nations with absolute advantages in so many resources were treated like desert-ridden geographies, with no purchasing power. Despite the abundance of freshwater available in Brazil for aqua-export production, controlling interests from Europe and the United States

revolted whenever the water resources were directed for the people of
Brazil and the development of the South American country (import
substitution industrialization). Import substitution industrialization is
a way a country such as Brazil tries to decrease its dependency on
imports, such as European imports, and use its own resources and
water for local manufacturing and production. This causes currency
punishment, as we'll see with Brazil as our example.

Hyperinflation was thrust upon the Southern Cone as Europe and
North America pounded South American independence with various
trade policies. For example, when Brazilian economic policy benefited
the country, developed nations dried up the flow of foreign currency into
Brazil, leaving the country with excess cruzeiro (the former currency).
Having insufficient foreign reserves causes a balance-of-payments
crisis, overwhelming even water-rich nations, generating high inflation.
The currencies of South America, in particular the ABC nations, were
constantly pegged or replaced because of the destruction Europe and the
US achieved through currency manipulations.

It was not until multinational corporate insurgence and settlement
became a welcomed trend in the late 1990s that the currency stabilized
and the Brazilian *real* (its new currency) showed the value of water
in its aqua-export-GDP-currency relationship. Because of the natural
resources and the accompanying lavish water supplies embedded in the
currency, the *real* will consistently thrive. And so will the multinational
corporations (MNCs) with strategic business units in Brazil.

$$\lozenge > \$$$

According to writers Bartlett, Goshal, and Beamish, traditional
motivations for outsourcing were "the need to secure key supplies
. . . By the early part of this century, Standard Oil . . . [was] among
the largest of the emerging multinational enterprises. [The] important
trigger of internationalization was the desire to *access low-cost factors
of production . . . Competition focuses on price and therefore on cost.* This
trend activates the resource-seeking motive."

Oil and water are *almost* inseparable, and their commonality as a joint resource is too compelling to be ignored. They are locked into the international political economy. One floats on top of the other. Water, billions of gallons, is used to lift oil from the earth's pores. So in the world economy, water and oil are one. Those who control one *or* the other have unimaginable wealth. Those who control both have unmitigated power.

The relationship between the two has long been recognized—and coveted. Our dependence and very existence as an industrial center is dependent on water and fossil fuels. Industrial barons like the Rockefeller family used the globe as a chessboard to control them. Wars have throughout history been fought over the quest for them and natural resources like copper in Chile—vital to extracting oil and water. In modern times, nations battle for control of both. Conflicts in Southeast Asia and the Middle East should be scrutinized for securing aqua-exports more than to create humanitarian relief or instill democratic freedoms. Henry Kissinger would definitely agree.

Our tensions with China are easily explained through aqua-exports. Our interference in countries like Chile has been, at its deepest root, about control of water. Water-rich commodities are the blood of our economy, an economy steered from its inception by those who procured natural resources and fossil fuels. Since the middle of the nineteenth century, those who controlled fossil fuels and the means to extract them helped design a world economy dependent on them.

$$\Diamond > \$$$

The early auditions for automobile engines in the 1890s evoked a *very interesting* group of choices: experimentation with steam engines, internal combustion engines, and electric motors (oddly, a twenty-first-century success story, despite its clear use in the 1800s). Fuel choices for the Model T were gasoline, kerosene, and ethanol. The Model T originally had a switch on its dashboard for the choice between alcohol *or* gas.

Coincidentally, it was Prohibition that made ethanol an impossible fuel for the Model T, and Standard Oil's economies of scale also made gasoline the cheapest option. Anthropologist Randy Amici once told me a story about how the Rockefellers' close ties to Prohibition and women's suffrage eliminated alcohol-based fuels. He smiled as he described the empty artifact wine bottles discovered at the homes where the women held meetings for the suffrage movement. Regardless of who did what, we grew dependent on gasoline instead of grain-based, ethanol fuels.

John Davison Rockefeller was a technocrat: one who creates the flows of where money goes; an influencer, who shapes global waves years before they break civilization's shores. Technocrats are brilliant and able to see beyond the *now*. They shape the future.

Let's pause for a moment and consider something relevant to the topics at hand—conspiracy theories about who controls resources and their motivations. Let's start with the Sherman Antitrust Act of 1890, the federal government's attempt to break up monopolies by criminalizing manipulative behavior. Section 1 of the Sherman Antitrust Act declares, "Every contract, combination in the form of trust or otherwise, or conspiracy, in restraint of trade or commerce among the several states, or with foreign nations, is hereby declared to be illegal. Every person who shall make any such contract or engage in any such combination or conspiracy shall be deemed guilty of a felony."

Section 2 states, "Every person who shall monopolize, or attempt to monopolize, or combine or conspire with any person or persons, to monopolize any part of the trade or commerce among the several States, or with foreign nations, shall be deemed guilty of a felony."

Can we not see that the very words defining conspiracy are laws established to prevent corporations from conspiring against the people?

Standard Oil faced countless accusations of such conspiracy, including coercion regarding physical threats to shippers and producers, utilizing shell companies, bribing opposing companies' employees to commit espionage, and restraining trade and promoting an outright oil monopoly.

Since 1890, "whistleblowers" who draw attention to corporations violating the Antitrust Act have sometimes been dismissed as "conspiracy theorists." Even the term *whistleblower* has a sometimes-negative connotation. Powerful yet quiet voices dismiss and stigmatize individuals who question corporate ethics. Please consider that "the powers that be" can make black sheep out of honest masses, journalists, and professors—through ownership of media and sponsorship at universities.

Among John Davison Rockefeller's contemporaries was President Theodore Roosevelt, the ultimate conspiracy theorist, who made it his mission to break up big corporate monopolies. South America intrigued Roosevelt, who recognized its tremendous natural resources. He took a dangerous journey and dark tour into the Brazilian rainforest along the Amazon—a massive river comprising almost 15 percent of the world's freshwater. Roosevelt wanted stability in that part of the world, recognizing the tremendous trade opportunities.

Other political moves related to water and oil seem far more sinister. In his book, *The CIA's Greatest Hits*, Mark Zepezauer reveals covert relationships during the twentieth-century that led to the formation of the Central Intelligence Agency—and the CIA's pivotal role in South America. When studying the history of oil, you discover, again and again, that wars and territorial disputes erupt more often over control of resources than over human rights.

In the 1940s, many Nazi figures saw their war efforts slowly deteriorate and power slide. There were negotiations between the US and these Nazi leaders because the US feared the USSR's future power. In 1943, future CIA Director Allen Dulles was sent to Switzerland to exchange immunities for information about US business partnerships with Nazi leaders. Dulles had been a Wall Street attorney with many clients, including Standard Oil. It is widely accepted that the Rockefeller family were world actors and technocrats who sourced and supported both US and Nazi interests. Dulles was also a secret agent for the Overseas Secret Service, commonly known as the OSS, and built foundations for the future CIA. He coordinated niche outside employment opportunities, which contributed to the CIA's success, created by the National Security Act of 1947.

These employees included General Reinhard Gehlen, one of Hitler's intelligence chiefs; Otto von Bolschwing, the Holocaust mastermind who worked closely with Adolf Eichmann; Klaus Barbie, ex-Gestapo chief who later arrived in Bolivia connected to the CIA; and Martin Bormann, Hitler's second-in-command. Zepezauer even mentioned how Martin Bormann may have faked his own death and found new footing in Latin America following the war, where he continually linked himself with the CIA. Numerous other accounts of the CIA's presence in South America, as well as migration of members of the Nazi Party, are well known.

◌>$

Following the war in the early 1950s, US interests helped to industrialize the economies of the so-called Asian Tigers: Hong Kong, Singapore, South Korea, and Taiwan. At the same time, high-grade iron ore was discovered in Latin America—ore that would help fuel the global industrial machine. The demand for rapid exports as prescribed to Asian Tiger economies was fast-funded by foreign direct investment. As a result, Asian Tigers were characterized by high GDPs, and much like the South American model, import substitution industrialization was not welcomed, forcing globalization. However, as we remember Brazilian tales, *these trade policies were not exactly free*, although they were maybe less extreme than in Latin America. South Korea benefited, emerging as a giant exporter in such industries as automobiles, chemicals, and steel during the 1970s. It became an exporter of vehicles in the 1980s. All of these are aqua-exports—extremely water-intensive goods and services.

Due to "growing public awareness and concern for controlling water pollution," the EPA introduced the Clean Water Act of 1972 and the Safe Drinking Water Act of 1974 that followed—both of which hindered corporate production in the United States. At that time, negative externalities such as corporate pollution were at a high for US citizens.

Asian Tigers developed *after* the United States military visited Asia. Could it be that foreign direct investment and high GDP from exports were not strictly correlated with *free trade policies*? Instead, perhaps US intervention allowed the framework for MNCs to enter and take advantage of extremely low-cost, unregulated, and abundant freshwater supplies? Krugman and Obstfeld offered an example: "Both the exports and the imports of Thailand soared in the 1990s. Why? Because the country became a favorite production site for multinational companies."

And so, corporations got in line like communists to invest in post-intervention nations, in pursuit of their unfathomable water supplies. First, it was foreign direct investments, then asset purchases, and finally citizenships, a.k.a. subsidiaries or strategic business units. Following military interventions, the US indirectly exploited these new industrial nations it helped spawn, by importing their cheap water through aqua-exports instead of using the United States' resources. Harmoniously, in the 1970s the US became predominantly service-based while the output of water-intensive manufacturing declined and industry slowed at home. The transition was in play for other countries to produce developed nations' needed goods. And as that played out, the US sank its hooks more deeply into South America, perhaps even orchestrating overthrows or revolts.

A *coup d'état* in Chile on September 11, 1973, that installed a US-friendly military regime later led to a "free market miracle" in the base of the Southern Cone. This event could be considered a hallmark of the development of Latin America's own Asian Tigers.

> The impact of free-market ideas on the Chilean economy and society deserves close scrutiny not only because Chile over a long period has been a free-

market laboratory, but, more importantly, because free-market advocates have pronounced the results a shining success. Since the mid-1980s, top officials of the World Bank, The International Monetary Fund, and the US government, influential US media, multinational corporation executives, and prominent neo-liberal economists such as Milton Friedman, have celebrated Chile's free-market experiment. They hold up Chile as a development model for nations around the world, from Russia to India, from Ghana to Bolivia. Chile has even been catapulted into the ranks of the Asian Tigers.

Joseph Collins and John Lear, *Chile's Free-Market Miracle: A Second Look*, 1995

While the CIA and well-connected US multinational corporations were already interfering in elections, radio, and other medians of propaganda in Chile, the overthrow and assassination of President Salvador Allende and installation of General Augusto Pinochet were more than Cold War chess during the times. The US Senate later reported: "The effort was massive. Eight million dollars was spent in three years between the 1970 election and the military coup in September 1973." Years that followed such assistance recorded Holocaust-like events in Chile. While thousands of sources do not deny US involvement, why was Chile a source of so much US intervention and US government support?

The US Senate Staff Report found that International Telephone and Telegraph (ITT) was the "most prominent and public example" of multinational corporate participation in Chile. The CIA/ITT relationship is only a paradox until you realize, after years of growth and asset sales, ITT's strongest affiliation in the international markets is water: it is the world's largest supplier of pumps and systems to transport, treat, and control water. *Fortune* magazine in 1975 disclosed that ITT was the tenth largest asset of all the stock holdings for the Rockefeller empire. It is hard to convey just how invasive the power and wealth of the Rockefeller Family were in the entire United States economy.

◊>$

The US interest in Chile seemed purely economic, *not political.* US interest fortified a Chilean government that would be a willing economic partner—exporting its valuable aqua-rich resources to feed the American dream to its north. Among its greatest natural assets is copper, the only noncorrosive metal that both oil extraction and water deliverance greatly depend upon. Copper reserves in the north of Chile are still documented as some of the world's best.

During my unrelenting journey for waconomics, I had to get closer to the corporations tied to water. I wanted to get inside the boardrooms. Quad and Tamaya Chemical's connection to Chile and Chilean copper and iodine was the reason I went to work for the organization—to back-test my theories for water-related stories. I was filling the roles of international business developer, chemical commodity trader, and analyst. I sat in meetings with CODELCO, a copper mining corporation connected to one of the "reasons" the coup took place. Moreover, the industries of copper and iodine had everything to do with water, and the success of Quad and Tamaya's CEO was catapulted because General Augusto Pinochet was voted out of office in 1988. Pinochet left behind a legacy of war crimes and mining rights in Chilean banks for aqua-exports like iodine, among many other opportunities.

In the 1990s, my president at Tamaya helped develop the Cosayach iodine mine while working for SQM. Fast forward to September 2011, and Cosayach was exporting *10 percent* of the world's iodine. That same year, it was discovered that Cosayach's mine, Negreiros, was stealing its water. The Chilean environmental authority demanded Cosayach halt the use of the water. Nothing happened. One month later, the Chilean government shut down the water well. In response, the price of iodine doubled in weeks. Just like that, one water well in the north of Chile got cut off, turning a 1.5-billion-dollar industry into a three-billion-dollar industry overnight. *Ten percent of the globe's iodine was temporarily removed from the world's supply because Cosayach lacked access to water, and the price of iodine increased 100 percent.* This affected many businesses and jobs worldwide.

Therefore, South America's connection and our vulnerability to water become an even more compelling argument for waconomics.

◊>$

In 1569, Gerhardus Mercator drew a map of the world with the equator much deeper south, which disproportionately oversized the Northern Hemisphere. The map of the North was intentionally drawn then as a giant, a colossus to be feared by everyone to the south. Peter Winn explained this in his *Americas*:

> "It is all a question of perspective," a Chilean philosopher explained to me in 1972, as he spread the map of the hemisphere on the sand near the crashing waves of the Pacific. "Like you, we have been accustomed to viewing our country from the north, and feeling our isolation and subordination as 'Chile,' which means 'the land at the end of the world' in [the] language of our Mapuche Indians. But, once you realize that our north is south," he exclaimed, turning the map on its head, "everything changes! Suddenly, Chile is at the top of the world, not the bottom."

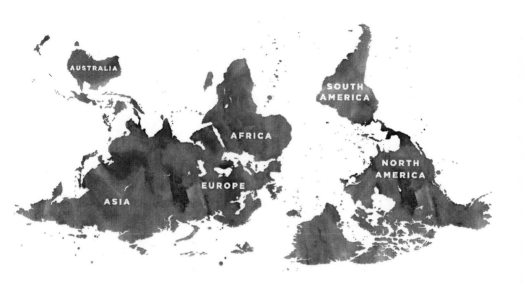

The truth is that South America is massive. The Andean Mountain range scrapes the skies at over twenty thousand feet above sea level, which makes the Rockies a less "lofty extension" of the Pacific spine. The river melt that leaves this range feeds the Amazon, Orinoco, and Plata rivers. Winn went on to say, "The Amazon is the world's mightiest river . . . and flows through its largest rainforest, covering 2.7 million square miles of South America, an area equal to 90 percent of the contiguous United States."

While the Amazon rainforest may be considered the "lungs of the earth," *water*-richness makes this abundance possible. Chile, in a world of consumer-driven desertification, with man increasingly relying on water to produce every resource imaginable, is clearly positioned in the ranks as "the land at the end of the world." The source of the water begins in Chile as it dries up in the Middle East—water wars are everywhere, it seems.

While "so many water wars are presented as religious wars," according to *Blue Gold: World Water Wars*, it is clear to see why wars emerge in what was once the Fertile Crescent. Lester Brown brings light to the reasons Israel has its enemies: "The Israeli population is roughly double that of the Palestinians, but it gets seven times as much water. As others have noted, peace in the region depends on a more equitable distribution of the region's water."

In *The Ends of the Earth*, famed author Robert Kaplan suggested that the pyramids were built with the Nile as the incentive, and total power was obtained through hydraulic societies. It was about control.

Chile and South America's importance will only grow, because industrialists recognize now, as before, that they need water—for processes such as hydraulic fracturing for fossil fuels. The Rockefeller family, for example, knew that if they owned oil, they must also own water. When oil became less tangible at the surface after years of extraction, water flood injection and hydraulic fracturing innovated access to reserves deeper into the earth's core, as shallower deposits dwindled. Eventually, the Rockefellers needed to *give water a value* or indirectly give oil a higher price tag.

In 1954, Guatemala became the first *democratically elected* Latin American government, falling to US-sponsored Cold War warriors, with the Rockefellers' influence playing a part. President Jacobo Arbenz was overthrown. Arbenz had appropriated unused land from the Rockefeller-owned United Fruit territory for land reform. He politely offered the listed price, but conveniently marked it down to one-fifth its actual value for tax purposes. Arbenz was bidding against the US ambassador to the United Nations, Henry Lodge, who was also a large stockholder and long-time advisor to the United Fruit Company. Secretary of State John Foster Dulles, and his brother Allen Dulles, director of the CIA, teamed up and launched unmarked CIA air raids and a rebel army trained on the United Fruit plantation in Honduras to replace Arbenz with General Castillo Armas. The oft-mentioned Allen Dulles, also Standard Oil's attorney, was thrilled to take the job. It resulted in a $20 million cost to the US, whereby three hundred mercenaries terrorized oil supplies and modes of logistics. More blood for oil.

◊>$

In response to water scarcity, oil companies have begun a different approach to production, yet their strategy has a questionable onus on supply. According to the Motley Fool, "Biggest, strongest, most efficient, most evil—there's hardly a superlative that hasn't been said about this most successful of the Standard Oil grandchildren," Exxon. And according to Hodge, "In a very secretive way, Big Oil has been buying rights that guarantee them access to freshwater in the future. In some cases, these rights give them access to water even before citizens and farmers. In Denver, for example, Big Oil has secured enough future rights to supply the entire city with water for three years."

Hodge went on later to say, "Big Oil's got the right idea. They're scooping up water rights on the cheap, hoping to sell them later when shortages become acute and the price rises sharply. It's an absolute right to profit." Blatant financial commentators *were not* around in 2006 when the idea of this book developed. Yet Hodge concluded, "The major oil companies are swooping up water rights at an alarming pace—before farmers and residential communities can get to them." Therefore, burning water through oil first may be portrayed as an intentional process—leveraging one need against another, burning the candle of profits at both ends.

The Rockefellers burned oil, hence burning societies' needs for water exponentially. Burning water in the twenty-first century seems intentional. Today's technocrat, T. Boone Pickens, has invested in both oil and gas and water rights separately. Ask T. Boone if he would be willing to contaminate his aquifers for fracking purposes. Owning millions and millions of shares in natural gas companies, Mr. Pickens inadvertently devalues a water supply during O&G exploration; hence, he drives up the value of his water rights elsewhere. The candle of profits burns at both ends. At one time, the car could only consume refined petroleum (aqua-exports), produced by water-intensive steel (an aqua-export), which is heavy and burns more oil. Then that creates a by-product pollution, which in sum destabilizes water quality and water use. While oil monopolizes energy, it does so more to our freshwater supply. And the car industry was obedient to the Rockefeller dream.

"In the area of fuel consumption, when asked in 1958 what steps his division was taking in fuel economy, the general manager of GM's Buick division quipped, 'We're helping the gas companies, the same as our competitors,'" according to Brock.

Big Oil wants you to use more diesel and gasoline. This burns more emissions. This indirectly creates the need for more water. So as the price of aqua-exports rises, diesel and gasoline will push water to the brink and create profit from its dwindling supply. It's an insidious plot to burn water and then raise its price as it becomes scarcer. Some countries have fallen into this trap, blindly or intentionally.

Because France was the first country to privatize water over 150 years ago, the model is French by design. Maude Barlow, author of *Blue Gold: The Battle Against Corporate Theft of the World's Water*, suggested that Margaret Thatcher, who privatized London water in 1980, initiated the privatization process. This is incorrect according to research. The US had already laid the foundation for covert water redirection. South American intervention was the real squeaky wheel. Author Ronaldo Munck discerned how England's policies may have been inspired by the "Chicago Boys" and a concert of Washington efforts led by the CIA:

> In 1973, General Pinochet was bombarding government house in Chile, putting an end to Salvador Allende's experiment in socialist democracy (as well as his life), and also launching a new economic model which was to have significant international impact. The Chilean economic model engineered by the so-called "Chicago Boys" (Chilean economists trained at the University of Chicago) was the right wing's answer to dependency theory and was to inspire some people in West, for

example Margaret Thatcher. The "Chicago Boys" acted
as a key link between the international capitalist elites
and the Chilean military project to carry out a veritable
"capitalist revolution."

J.D. Rockefeller founded the University of Chicago. Dr. Milton
Friedman, professor of the Chicago School of Economics, was also an
economic advisor to Presidents Richard Nixon and Ronald Reagan.
Prior to his tenure on Nixon's payroll as national security advisor,
secretary of state, and "Assistant President of the United States,"
Henry Kissinger was a known "Rockefeller man," a cabinet member
who would move societies for the family.

Kissinger said it best: "Control oil and you control nations; control
food and you control the people." But today, if someone controls water,
they can define the pain in so many ways. The United States (like
Europe) could not control water directly during Kissinger's reign, so it
first controlled aqua-exports via covert operations (military or political
intervention), then trade policy, next foreign direct investment with US
corporations, and ultimately privatization of emerging markets' water
supply through corporate overhaul. Kissinger's contemporary and a
favorite US economist, Milton Friedman served America best with *what
he called "Freshwater Keynesian" policies* in the 1970s. This name is no
coincidence. Rather, the obvious name is the cutest red flag in association
with waconomics. Friedman was calling it exactly what it was and still
very few realized water's connection to our financial freedom.

Kissinger spent the bulk of his time as secretary of state trying
to stabilize regions vital to American industrial barons—areas heavily
rich with water, such as Vietnam and Thailand, or water-dry areas with
an abundance of oil, like the Middle East. When we realize Exxon
ended up drilling for oil in Vietnam, all these connections tighten.
Those efforts later associated the secretary of state with war crimes. In
the Middle East, it would be very difficult to extract oil if water wars
were constantly interfering with oil production. It makes more sense

to control water wars by covering them up and by making another war elsewhere in that country or region—redirecting destruction to where it is least relevant to oil production.

In other parts of the Middle East, MNCs want to produce oil and gas from a region, so they stabilize a country by subsidizing its water supply. Qatar is an example of the paradox of value. *The richest nation per capita* is a country well below the water poverty line, with only forty-eight hours of its own emergency water supply. Qatar is also known to have one of the highest water usages per capita to boot. How is that possible? Qatar, mainly a desert, imports 90 percent of its food, an aqua-export, and exports energy, another aqua-export. This simple exchange of goods and services makes sense in theory, but where is Qatar getting water to extract and produce its energy?

Qatar's energy, which we can consider as aqua-export advantage, invites creative investment in the country's water supply through desalination and oil prices. Corporations subsidize desalination plants to maintain access to Qatar's fossil fuel reserves. It's all a subsidy with fossil fuels, one way or another—including free water and healthcare for the people. Qatar's abundant reserves of oil and natural gas (O&G) and its location as a peninsula-port on the Persian Gulf allow it to leverage water supplies for its very valuable fossil fuels. Please note that desalination is extremely energy-intensive (first toll) and thus extremely water-intensive (second toll). In other words, it is extremely expensive, requiring you to pay the toll twice to use an unstable bridge once. The fossil fuels required to power desalination plants are finite in Qatar. Once they are depleted, will Qatar still have access to water?

Qatar may prove insulated and protected from water wars within the interiors of the Middle East because of this aqua-export advantage. For now, Qatar experiences a rare stability in the region. The same is *not* true for Yemen.

Sana'a, in Yemen—one of the Middle East's terrorist hotspots— has long been projected to run out of water as early as 2017, making it the world's first capital city to do so. The water situation is so serious that the government has considered moving the capital. Headlines in 2017 certainly convey rising panic. However, those troubled stories are without specified connections to water. What happens to a nation's currency once its water piggy bank is all used up? We should investigate the economic weakness of Yemen and how water affects its currency. The Yemeni *rial* is so water-less that it has substantially fallen in value against the US dollar throughout the first decade of the twenty-first century. The rial is one of very few currencies to do so during this period. With all of the dollar's flaws in that decade, how was this possible? Yes, there are media articles suggesting the disarray of the Yemen banking sector, but the banking mess is a consequence of water supplies. Since the rial is not the world's reserve currency like the US dollar, and since Yemen does not have any aqua-exports like Qatar's oil and gas to prop it up, when water is short for people and production, it simply cannot save its currency or industries like banking. It's all connected to water one way or another. The economic and political volatility racking Yemen is water-related.

We must begin to ask ourselves, are terrorism and war promoted in regions that lack enough water for the people? If we reconsider all the areas in the world where religion is predominantly *the reason* for war, isn't it overwhelmingly consistent how readily available water supplies are *almost always missing* from those people and that country?

CHAPTER 9

History Repeats Itself ◊

DEVELOPING COUNTRIES CATCHING UP to the powers of Europe, the US, and China are paying ten times more for water. Simultaneously, these poorer nations receive significantly lower wages in weaker currency. The water gap is often widest in these underdeveloped nations of the world. We, the fortunate, indirectly take their water.

Just like Big Oil can take water rights in Colorado, it's rumored that other individuals—the Bush family—took hundreds of thousands of acres in Paraguay. According to *Blue Gold: World Water Wars*, if we include the first *and* the second President Bush, 185,000 acres to be exact. Validated in 2006, *The Guardian* described 100,000 acres of land "allegedly" purchased by President George W. Bush during his daughter's official visit to Paraguay.

The catch: this is the area above the biggest natural aquifer in the world—the Guaraní Aquifer, just a downstream neighbor from Chile—in a neighborhood set up by years of intervention.

The road toward water privatization in Chile began on September 11 in 1973, through a military insurgence invited by the US after a decade of efforts. During the day of the coup, a 500-page economic game plan designed by the Chicago Boys was placed on the desk of each general serving the new military regime by twelve o'clock that next night.

"Sweeping macroeconomic reforms" took place just weeks after September 11, 1973, "including privatization, price liberalization and the freeze of wages," according to Canadian author Michel Chossudovsky. He went on to say that just weeks after the coup, "the military Junta headed by General Augusto Pinochet ordered a hike in the price of bread from 11 to 40 escudos." Price hikes in bread alone increased 264 percent overnight, attributed to the "economic shock treatment" designed by the Chicago Boys. "From one day to the next, an entire country had been precipitated into abysmal poverty; in less than a year the price of bread in Chile increased thirty-six-fold (3700 percent). Eighty-five percent of the Chilean population had been driven below the poverty line . . . wages had been frozen to ensure economic stability and stave off inflationary pressures." Currency devaluation and price liberation hiked the price of imports that Chile badly needed, causing more inflation, making food and fuel all but impossible to acquire. Admiral Merino, a member of the Junta, warned, "We will be accused of killing the people with hunger."

This was a *false* inflation pulled from that 500-page economic game plan. Before the coup, Chile was known as one of South America's soundest nations. Politically stable and democratic Chile was comparatively advantaged and well-balanced in so much. These prices sent contradictory CliffsNotes to those bandwagoners who later appraised Chile as a "free market miracle." Hyperinflation was an engineered price, not the natural temperament of one of the continent's most efficient economies. The setup unfolded in Chile as tariffs were lowered, predatory pricing occurred, and dumping raised prices. Predatory pricing, or price discrimination, is "the practice of charging different customers different prices," and dumping is "a pricing practice which a firm charges a lower price for exported goods than it does for the same goods sold domestically," according to Krugman and

Obstfeld. And connectedly, aqua-exports became drastically expensive for Chileans and relatively cheap for the United States—rampantly increasing the foreign demand for Chilean goods. This fed the standard of living for the US in the long run. It was a substitution of natural resources and low-cost factors of production for the US, in exchange for increased South American poverty.

The "free market-miracle" was led by something not so free: a peso pegged to the US dollar in 1979. The peso was pegged at the riskiest time relative to US dollar history—meaning the dollar was very oversold, and when the dollar eventually bounced, everything locked in at that peg would get expensive. It was a bear trap. This was an environment in the mid-to-late 1970s which created excessive import consumption—essentially buying dollar-depreciated exports and loans from the US. This led to artificial Chilean growth—discounted US imports and loans, which ventured off from vast private borrowing. It also caused tripling of foreign debt from 1975 to 1982—some of the most extreme in South America. At the time, homes in Miami were cheaper than in Santiago. This peg in hindsight seems like a pragmatic foundation for debt and, constructively, foreign control. Chile's economy was set up and framed.

A bubble then developed, which later paralleled the "Great Recession" in the United States. President Reagan's monetarist policies, tenets of Chicago Boy economics, encouraged it. During this consumer's paradise, the economy appeared healthy. On the 1980 anniversary of the Chilean coup d'état, a new constitution won 67 percent approval in a carefully controlled plebiscite.

An example of this chartered progress was the "system of tradable water rights" developed in 1981, which offers a recommended manuscript for developing countries. As historians Mónica Ríos and Jorge Quiroz described the view, "A private market in tradable water rights would maximize the economic value of the resource, would help to reduce costly public infrastructure investment, and would foster private investment in irrigation." Private investment in irrigation

triggers thoughts of California's water and its "Wild
overdrafting—taking water and the unknowns about i
were taking.

Due to unequivocal inflation in the US at that time, F.   ..nan
Paul Volker's huge spike in interest rates caused a fall in commodity
prices and, in lockstep, a dollar appreciation. Chile's debts were tied to
US interest rates and the US dollar. A pegged peso meant that if US
rates and US dollars rose, then Chile's debts rose. This precipitated
further indebtedness to any country linked to the dollar and further
increased the real value of debt and its mounting service requirement.
This sounds like a rerun of the United States' housing bubble at the
beginning of the twenty-first century, except that the plays from this
playbook happened in Chile first. Just like the "Too Big to Fail" US
banks responsible for predatory loans needed bailouts by the US
Government to stay alive in 2008 and 2009, Chile, too, needed bailouts.
The Latin American Debt Crisis arrived in 1982, inspired by surging
US interest rates.

As Collins and Lear described it, "The government wound up
taking on as public debt some $16 billion in foreign loans, most of
which had been originally incurred and often recklessly spent by private
Chilean conglomerates." So the period where the peso was linked to
the dollar during Chicago-drawn US inflation was then drastically
devalued (Milton Friedman was a close advisor to Ronald Reagan, and
therefore Paul Volker). This increased the Chilean debt for consumers
by "70 percent in peso terms." And "in 1982 the government ended
all indexing of wages and salaries for inflation, and the Labor Code
decreed in 1979 severely restricted protests and collective bargaining."
It was no wonder that even supporters of the decade-old coup now
staunchly protested the current government policies. "One out of eight
Chilean workers wound up in government emergency work programs,
typically jobs planting trees or sweeping the streets and at less than the
miserably low minimum wage."

Under that regime, Chile's debt grew and reached its high in 1985, at $21 billion, 100 percent of its GDP, forcing export-led growth to meet its new needs for loan repayments. In between 1982 and 1988, the peso was pummeled, and depreciation rocked Chile's money. The currency was in free fall and the discount on water-rich aqua-exports leaving Chile was a fire sale to buyers. Imports were noticeably reduced with prohibitive tariffs, and price discrimination occurred in its domestic markets. Boiled down, this priced the Chilean people out of their own aqua-exports!

Collins and Lear described the oft-mentioned *noose*: "The Pinochet government's dutiful payments to the foreign banks and its persistence in free-market reforms (by this point strongly championed by Reagan in the US and Thatcher in Britain) won Chile the favor of the IMF and World Bank, which funneled short-term credits to help pay interest on the debt and supervised the financial restructuring."

After the government had *already* sold companies—assets of Chile's before the coup—which had failed during the 1982 debt crisis, the "Chicago Boys launched a second wave of privatizations that included Chile's largest traditional public companies, such as electric utilities and communications monopolies." These companies were sold at ridiculous discounts. Meanwhile, as Collins and Lear continued, "the huge debt Chile took on under the Chicago Boys' watch became the instrument by which many important national assets passed into the hands of foreign companies." Via debt-equity swaps, foreign companies enjoyed the purchase of natural resources at about 30 percent below market value. In sum, Chilean citizens were losing.

The "free-market miracle" of the 1980s was "based on an expansion of exports, particularly such non-traditional exports as fresh fruits, seafood, and forestry products," a.k.a. new shipments of aqua-exports. *Under Pinochet, Chilean fruit exports boomed.* "Shipments overseas of grapes, nectarines, plums, peaches, pears, and apples grew at a compound rate of nearly 20 percent annually over 18 years, from about $40 million in 1974 to nearly $1 billion in 1991 . . . about half of all of Chile's fruit exports go to the US." And "the fruit industry has also been able to count on a supply of abundant water whenever needed, thanks to good sources of water and an extensive network of irrigation canals." Here, Collins and Lear referenced the tradable water rights developed in 1981. Chile's forestry was being mined like copper and exploited in the short run with no sustainable long-run vision of renewal. This resulted in a dangerous mix of damaging watershed, pollution, and deforestation.

It is said that a tree standing is worth three times as much as a tree cut, because the root system holds the soil and the water enriches the fertility around it all. Once we remove that tree, we remove the high-quality soil content because we lose the ability for the land to hold water. During the free-market experiment in Chile, there were basically no environmental penalties for destruction to the land or water. Everything done to the environment was a social cost and unrecognized leak (UL). Growers, farmers, and corporations could do in Chile what they *could not* in the United States after the EPA. Chile absorbed pesticides like Agent Orange and 1080 poison to enhance productivity; Chilean practices of river contamination and air pollution are both well-documented. Chile had an environmental case against its privatized Chicago-styled growth and "free market miracle." Unbeknownst to Chileans and North Americans, Chile's environment was exploited with practices the EPA wouldn't allow, and we all absorbed that poison. Guess who loses there?

The US ambassador to Chile, Curtis W. Kamman, said in May of 1992, "I challenge you to find another country in the world that has such nice stats. It's almost too good to be true, and people wonder where the downside is."

Not so fast. During this free-market era in Latin America, wages decreased while poverty climbed like a tortoise to the road. From 1980 to 2004, those living in poverty increased to a twenty-four-year high. The macroeconomic picture of the global structure had been rebalanced for profitable intent. Still unsatisfied, the World Bank in 1997 confessed that it was "all pain, no gain" in Latin America. Krugman and Obstfeld agreed: "Growth rates in Brazil and other Latin American countries have actually been slower since the trade liberalization of the late 1980s than they were during import-substituting industrialization." Remember that import substitution industrialization is a way a country tries to decrease its dependency on foreign countries' imports and increase its own local production.

So domestic prices rose, wages decreased, and poverty climbed, yet huge investment was pouring into these countries—results that seem so counterintuitive.

More coups or military intervention occurred elsewhere to implement the strategies experimented in Chile. Simply put, multinational corporations got in line like communists for the rights to exploit developing nations and then sell the goods to their home country as if those prices were based on "free"-market principles. The fruits and the fruit prices sold to the United States were not the results of pure capitalism; they were socialized prices because of intervention. The technocrats of the world wanted countries with the most resources to be indebted to these international institutions. This is "beggar-thy-neighbor" policy. But with water, it's so much more personal.

Water constitutes life, health, and growth opportunities. Chile's acceptance has allowed other nations to move forward in a climate where foreign direct investment is vast and genuine public opinion is harsh. "In due course, the economic bullets of the free market system were hitting country after country. Since the onslaught of the Latin American debt crisis, the same IMF economic medicine has routinely been applied in more than 150 developing countries," wrote Chossudovsky. This "free-market miracle" was not free of government intervention; it was cyclical socialism, again and again.

"The 1980s saw the rise of the 'technocrat' as manager of the nation's destiny in much of Latin America. Society had become, to some extent, depoliticized, and decision-making would be based on 'technical' criteria. At this time, the state was selling off many of the state enterprises and ceding many of its functions to market mechanisms. The promise to the population was that consumption would now take the place of politics," Munck wrote.

The technocrat, according to Munck's research, was bolstered and fortified by the IMF, United Nations, and World Bank during these days. "The 755 sales and transfers to the private sector that occurred in Latin America between 1988 and 1995 represented fully half of the value of all privatizations occurring in the developing world. Nearly half of these sales were in the area of utilities where the possible gains were seen to be greatest." Utilities, municipalities, waterworks—the continent with 31 percent of the earth's freshwater was getting bought by foreign governments, all while environmental protections and regulation were screaming higher in Europe and the United States.

> The privatization of Chile's largest water companies began in late 1998, a process prompted by the rising demand for sewage treatment, for which the government claimed to lack funding. Three out of the five privatized utilities were awarded to U.K. water corporations, while French and Spanish investors won the remaining two. Within three years, private water companies were servicing more than three-quarters of Chilean households. In Chile, largest urban centers such as Santiago, Valparaiso, and Concepción, private water companies cover nearly 100 percent of the service.
>
> Food and Water Watch

Chile is now a privatization-friendly country in the Southern Cone, allowing the distribution of its most precious resource to be governed by corporations for private profit.

Ultimately, the efforts of the Chicago Boys created a boom for "water" real estate.

According to Michelle Caruso-Cabrera, "Chile takes such a commercialized view of water, it spawned a new occupation, water real estate agents." CNBC thought it was worth covering *Chile's water story* with the American people in 2010. Chilean water real estate agent Sandra Vilches-Brevis stated, "In Chile, if you own a piece of land that has water on it, you can sell the rights to that water without selling the land itself. Conversely, you can buy access to the water without buying the land."

When CNBC's reporter asked how much the price of water had climbed in the last ten years, Vilches-Brevis explained that it had almost doubled.

"It's doubled!" the reporter repeated in disbelief.

In the game of Monopoly, The Water Edition, everything is Boardwalk and Park Place. The Chicago Boys were so inconsistent with their free-market approach that the only apparent impression was extraction for profits—what we can call the Chilean people's water gap—generated by social costs and ULs. The problems were engineered for the creation of this purpose. This was what Collins and Lear call the process of "privatizing what was valuable and socializing the costs."

President George Bush, Sr. spoke in Chile in December of 1990: "Chile has moved farther, faster than any other nation in South America toward real free-market reform. The payoff is evident to all: seven straight years of economic growth ... You deserve your reputation as an economic model for other countries in the region and in the world. Your commitment to market-based solutions inspires the hemisphere."

The question worth asking is, what did privatization do for the people of Chile? Maybe we should compare country notes and see if history repeats itself.

$$\lozenge > \$$$

While the Chilean performance was not the first successful CIA-led coup in South America, it was applauded over the years despite its publicly questioned win. Bolivia's story is not dissimilar to Chile's. The CIA cookie-cut the process, and the military intervention constructed the execution—we can discuss the Southern Cone or go elsewhere to prove this standard operating procedure. Yes, suffocating debt is a means to an end—control. Unfortunately, the scheme is a debt-service payment mediated by international groups such as the IMF or World Bank in exchange for a country's assets. This traditionally comes at the expense of the children, countrymen, and women. A true third-party event, filmed for the world to witness, occurred in Bolivia. SUEZ, one of the world's largest water, electric, and wastewater providers, the result of a 1997 merger between Compagnie de SUEZ and Lyonnaise des Eaux, installed its subsidiary Aguas del Illimani into El Alto, La Paz, Bolivia.

Jim Schultz, founder of the Democracy Center in Bolivia, shared the painful circumstances in which corporations gain the water rights through international instruments, as shown in the documentary movie *Flow:*

> Water privatization was absolutely forced on Bolivia by the World Bank. In 1997, the World Bank told Bolivia that if they did not privatize the water system, of Cochabamba and El Alto, La Paz, that they were going to be cut off, from water development models by the World Bank . . . How can you keep loaning poor countries money to meet their basic needs? It is completely predictable. That what happens is poor countries end up with debts they can't afford, and that debt becomes the noose around their neck with which the World Bank and the IMF become the government.

In Bolivia, water ended up being more expensive than food, denying 208,000 citizens their basic standard of living.

Ironic as it may be, *multinational corporations*—which have been themselves *subsidized* and *supported* by the World Bank and the IMF—*believe* water is no longer a *public good.* Chairman and past CEO of Nestlé Peter Brabeck-Letmathe has directly said, "The one opinion, which I think is extreme ... declaring water a public right. That means that as a human being you should have a right to water. That's an extreme solution."

Maybe you're right, Pete. It's something greater. We can demonstrate that water is a public duty, public responsibility, and an economic game changer that can return so much more in profits than privatization could achieve. We can pump water and all its strength toward total benefit. The returns are ours to be had, if we want them. Or we can continue to regress at home, where evidence is less glaring.

$$\text{\textcircled{\$\lozenge$}}>\$$$

When states turn their municipalities over to private firms, the people in the poorest countries with water abundance have lower income-to-water rates—household's income divided by a household's monthly water bill. They do not have enough money for the water fees imposed relative to all needs. Consider it like credit. Poverty needs credit but gets no support from the banks. Wealthier individuals do not need credit, but qualify for its access. It appears in developing countries that involuntary debt must come before credit—as suffocating from debt is then a riskier endeavor lured by intolerable interest rates. So big multinational institutions like the IMF, UN, and World Bank leverage their multinational interests with more tolerable loan structures for the poor.

Poor households, under privatization, "end up paying water prices that are as much as ten times higher than high income households," and "that average price of water declines as household income increases," according to Cristina David. This was the case in Manila, Philippines among Bechtel and SUEZ with the Metropolitan Waterworks and Sewage privatization.

The three largest water companies are known all over the world, but sometimes by different names. They are Vivendi Environment (also known as *Société des Eaux de Marseille, Compagnie Générale des Eaux,* or *Veolia)*, SUEZ *(Lionnaise Des Eaux)*, and RWE Thames. Talks of mergers have already begun. Their emerging role seems to be taking control of developing nations' water supplies. In 1990, about fifty-one million people got their water from private companies, according to water analysts. After its last update, that figure was then estimated at more than three hundred million.

These companies have even gained contracts in US cities such as Atlanta, Chicago, New York, Las Vegas, Pittsburgh, Tampa and many more. They rarely use their true names; for example, SUEZ equates to United Water in Atlanta, Georgia. "There isn't an example in the world where there hasn't been a hike in water rates" upon their arrival, according to *Blue Gold: World Water Wars*. After SUEZ took over water contracts in Atlanta, Georgia, Atlantans were boiling their water—and these are citizens in rich nations. We're going to need to remember this later . . .

"The evidence that markets work badly is less glaring in advanced countries," Krugman and Obstfeld wrote. What happens when SUEZ's model turns against the United States, slowly administering less glaring evidence at home?

JeanLuc Touly, the accountant of thirty years for Vivendi/Veolia Corporation, spoke in *FLOW* about the expertise he gained in their work environment.

> These companies date back 150 years. And they were created by bankers; this is what you need to know. These multinational corporations are by no means philanthropic organizations. So the progressive language of these multinationals is truly scandalous. Because they say, "We are the ones who will end poverty with regards to access to water..." But how could Vivendi shareholders wait 10 to 15 years to bring water to people who can't pay for it? They are not interested in that at all.

"What was being sought by the architects of the new economic model was a stable and profitable international division of labour led by the financial sector," wrote Munck. Bankers created all of this. They understood water was a better currency than gold in the desert, and that everyone and everything is predetermined by its use regardless.

Privatization in developing economies appears beggar-thy-neighbor. The strategic policy increases welfare for corporations and shareholders at the progression of underdevelopment globally. History tends to repeat itself, as was seen in Chile and Bolivia, and the laundry list of nations continues, while the greatest pattern and our biggest bang for the buck, water, is incrementally holding down everyone.

CHAPTER 10

Water Bubbles

EVERYONE SUFFERED ON DIFFERENT levels as the "Great Recession" approached—the year 2008 was on the game clock. Team USA was hurting and exhausted—government officials, the corporate partners, and we the people all reacted to the score of the global economy in different ways. In hindsight, it was smoke and mirrors on the TV. Nothing could slow down this game but a critical replay.

When we look at the 2008 highlights, it's like an entire basketball season throughout various arenas in a few different time zones. Government officials managed the economic game, like refs adjusting the clock. All the while, the Federal Reserve coaches looked dapper in their choice suits. The corporate partners were starters, and we the people came off the bench as second-stringers in this Great Recession. Because of how game-time decisions were made, the importance of time and each successive minute played favored the corporate partners. Eventually, it was you and I who came in for the exhausted and injured first-stringers. By that time, we really felt the pressure of the economic season and Team USA.

There was pain for all those involved, but each had a unique experience, role, sacrifice, and time zone. The Fed coaches, government refs, corporate starters, and we the people were all affected by the economic game clock differently. The Great Recession started many seasons before the actual pain ever materialized. And yet *everyone* chose the usual suspects for blame.

◊>$

The hills of Southern California raged with wildfires for what seemed like years. And the flames finally crept their way into the swamps of North Carolina where they smoked into the summer of 2008. Dust storms threatened the Olympic Games in Beijing, and the world felt dry as humanity congregated in a drought-stricken China. Economic cycles and national teams were overheating from the competition. Gas station lights flashed $4, $5, and higher as quickly as the weather fluxed. Replays from nightly announcers painted pictures of climate change from 2006 to 2008. During this period, the global economy experienced net drought and the earth was on fire—lacking sufficient water for all aqua-exports, all manufacturing. The unexpected knockout performance of persistent global droughts raged between the summers of 2006 and 2008 and left everyone in the international economy mesmerized by the symptoms.

The world was clearly short water because of Mother Nature, but the attention was mainly on oil—the headline. Perhaps it's easier to wrap our heads around the past and its financial consequences like loyal oil fans. We've held energy prices accountable before in the 1970s during the oil crises, rather than do the same for something basic like water, which influences it all. Was water responsible for all the symptomatic changes in 2008? What the US consumer heard was that oil was causing food prices to rise, and most accepted that to be the reason for life feeling tight. Later, a play-by-play coverage of the housing market took the spotlight.

Yeah, we blamed the houses, too.

The government officials, the corporate partners, and we the people all experienced different opportunity costs of this reality. Everyone's purchasing power was profoundly reduced.

The best example started on the West Coast of the US in 2007. What many do not realize is that while California was on fire, it had a quiet role as the third largest oil producer in the US—and it was short water, too. California generated 244 million barrels of oil per year, just shy of Alaska's 263 million. During such droughts, it needed 2.56 *billion* barrels of water to make that happen. And at the end of oil's production in 2007, California was contributing to 12 percent of the nation's produced water, a.k.a. "drip gas," and potentially growing our food with that hazardous waste. Essentially, if California used that drip gas to put out those wildfires, who really knows if those flames would dwindle, or how much fuel to the fire drip gas would add? During such a mega-drought, the people of the world depended on California's produce, compounding the global water shortage from the fifth largest supplier of food. Connectedly, devaluing 12 percent of the water supply for growing food required excessive dollars and time to treat that hazardous water after oil and gas production. Costs were rising to reintroduce water to humans and for humans to coexist with that devalued water. This was all happening as oil costs were rising. So what caused them to rise? Did oil move up in price because water was short for manufacturing it?

As oil prices rose, corporate America was being charged higher operating costs—we call them fuel service charges for transporting goods—and they were tacked onto the bill, causing everything to spike in price. Prices were moving too high, too fast. This eventually led to a collapse in all prices, a.k.a. demand destruction. During the stock market's plunge, we were told money went into a "black hole" and disappeared. But money didn't disappear: the world was charged for water and didn't know how to say it, or digest that overwhelming cost.

Water inflation hit the consumer's pocket hard, yet everyone believed things were getting expensive because of the price of oil. The stock market was falling and we the people were paying more for goods and services, a.k.a. water-intensive aqua-exports. People started to become more conscious consumers, using car pools and coupons in their grassroots effort to manage their personal buying power. Why would the US anticipate water causing the price of everything to rise and the stock markets to fall simultaneously? Who on the news stated that "all this" was water inflation and came from those swelling worldwide droughts and rapid fires? *Tonight on* 60 Minutes, *water inflation is ramping up to a heavy financial blow, which inevitably will bring the world to rethink its needs!*

The price of food rose in response to the lack of world water—not oil—and oil prices rose in response to that water shortage as well. Fuel service charges were just a domino reaction to it all. In economics, they say prices are sticky and food inflation takes a considerable amount of time to find its way into the consumer's pocket—and the actual perception of financial consequences. Therefore, food prices spiking before our eyes were something entirely different.

Ticking up from January of 2007 to July of 2008, oil became 191 percent more expensive—increasing from $51 per barrel to $147 per barrel. Since water was not yet a tradable commodity, we couldn't directly speculate on the cost of water for oil production and it found no place in the media other than the weather forecast. Oil, coffee, corn, copper, and timber are substitutes for water, a.k.a. aqua-exports, and they were traded in water's place. Day traders follow trends, and at that time agriculture, commodities, and energy were moving up quite nicely because water was scarce. Stock operators just traded aqua-exports in 2007 and 2008 as the vehicle to trade water—while 99.9 percent of them traded commodities without knowing about a water connection. The traders bet on farmers and producers of aqua-exports, who bring these raw materials to market. Other investors bought homes because they were rich with commodities, and thus they too rose with the price of aqua-exports. When water was short, the price of everything rose.

Use the chart herein to see the timing of it all.

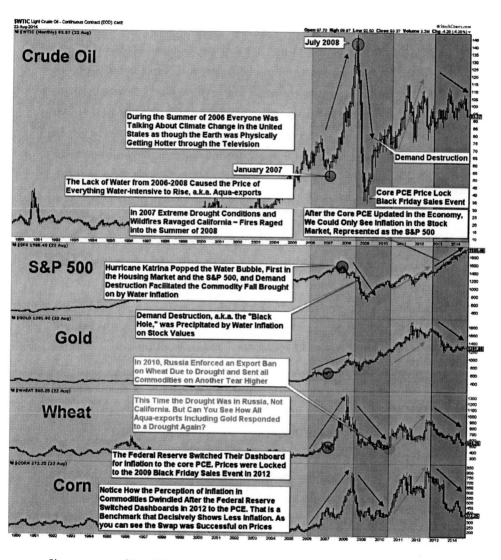

Chart courtesy of StockCharts.com

◊>$

Keeping up with the drama of price spikes and the shocking heist
of our buying power in 2008, water was simply not in the lineup of

suspects. Yet the usual candidates we gossiped about, oil and then food, all had the same loud and clear birthmark of water that no one paid attention to. At the time, I explained to my closest colleagues and family that prices of everything were rising because of water. While wa-conomic theories were not publicly stated, this market situation lit a wildfire under my ass to continue. If everyone paid attention only to the usual suspects, how would anyone step outside the box to consider water's shortage as the reason all commodities were moving up in price? More importantly, how could we turn all future pain into gains?

Economist James Roberts reflected on food prices moving prior to oil surges: "Since the fall of 2007 there have been food riots in over twenty different countries, most of them near the equator. The Prime Minister of Jamaica resigned in response to food protests. Twenty-two nations of the world have imposed export prohibitions on their food because they don't want to sell it to others when they are starving themselves."

Food is the most water-intensive of all commodities, sporting the biggest birthmark, but oil grabs everyone's attention and it became the frontrunner in the news. Aqua-exports took their cue from the lack of freshwater on earth and stamped the validity of *water value* in the world then, whether the world took notice of water's *voice* or not. Water value was a game changer.

We the people easily related to the shocking new price tag of energy, as it affected our everyday power bills and gas prices. For individuals, it was escalating the cost of a ride; for businesses, it was the fuel service charges. We did double takes at the sight of hikes in gas station lights, and nostalgically shared stories with each other of how cheap it used to be in the "good old days." The Prius was suddenly sexy and the Hummer II was never scarier. Then, from gas stations to grocery stores, we noticed some hikes in food prices, and never doubted it was because of the cost of transporting everything—a.k.a. oil prices. That's what the news told us. If the price of food didn't increase, then the size of the boxes and containers got smaller. We paid the same amount for food, but got less—and we noticed that, too.

Even today, there is less food in the same old box. For example, a bag of bagels contained six doughy delights for what seemed like decades. And as of 2017, an organic, non-GMO brand hits the shelves much higher in price, and *only contains five in a bag*. Obviously, *water inflation is here to stay*. Just like food prices are sticky and won't fall, the boxes never get bigger again for the same price. Retail prices we see at the grocery store or gas pump typically never fall back to previous pricing levels, they only pull back to higher lows and rise relative to history. That is why we say they are sticky—because prices never fall from the rising ceiling. During the same time period, many suspected that the Iraq War was for cheap oil. If we were subsidizing a war for cheap oil, then where was the payoff in gas prices?

All in all, it was easy to follow oil and its distinct role in our lives, rather than leaving our comfort zones to believe otherwise. But agricultural spikes that cause food price spikes, which last in grocery stores into perpetuity, are created by something extraordinary—not short-term pops in oil prices.

With everyone focused immediately on relieving the pain from the economic hit by oil and food prices, we all missed the big *X* that marks the spot—water. The symptoms had frozen and intoxicated us all. Consumers and businesses experience economic pain at different times, while the government creatively juggles perception. There are many different time zones in an economy, and the stock market always reacts first.

◊>$

Water in the stock market is like a martyr. It shows the world its dedication with no visible credit, other than its reflection in the prices of aqua-exports—a.k.a. the prices of goods and services corporations sell and turn into profits. In this sense, water can be viewed like the parents helping their children, aqua-exports, with a science project but taking no credit at the science fair.

Pure water could not be sold or bought on the stock exchange, just like parents are not officially allowed to enter and participate in science fair competitions. Instead, it entered the arena via commodities. Commodity speculators at that time probably did not realize water was directly affecting price, and many may still overlook such force today. It was, I believe, *the pent-up shortage of water between 2006 and 2008* that helped the markets crash, especially when we understand water's force was muted from interest rates, and Hurricane Katrina popped that housing bubble. So let's go back to the winter of 2007 and take a look at water's involvement during that period, watching the highs and lows of our players' actions.

When the equity markets, like bank stocks, began ramping up to a peak by that 2007 summer, a lack of water in the forms of droughts exacerbated shortages. Speculators bought the next best representation of water, which were commodities and the producers of commodities, a.k.a. aqua-exporters like Exxon. The trends in commodity markets roared like sudden climate change, and the stock values of those producers flew higher and peaked with the commodities in July of 2008. It was a great ride for those farmers and producers, until that summer when prices without a suspect caught up to them. *This was water inflation at work for them.* Producers of aqua-exports like food and oil felt the burden of prices moving too high too fast, eventually eating away at their bottom line, with prices such as fuel service charges cannibalizing their own demands.

The players in the economy felt reality weighing heavily on their reckless expectation in the demand for all commodities at those high prices—realities that pop bubbles. Such high prices drove out future buyers, and demand destruction occurred and the bubble on aqua-exports popped. Corporations love higher prices, but not when those prices self-destruct. In that circumstance, the corporations are left holding the financial "bag," trying to sell goods as everyone chases prices lower. Finally, everything in the stock market started collapsing this time, including commodities and commodity producers. The "black hole" where money went came from all those additional costs that water inflation caused. With a shift in perspective, we clearly see how the lack of water could cause havoc and cannibalize costs for Team USA. It's not the price of oil alone which causes the cost of transports to rise, food prices to increase, and food boxes to shrink. It's water first.

No matter what, water remains consistent in its impact, regardless of how we try to interpret the stock market's behavior. And so does demand destruction, which, as we will later learn, hurts water more than anyone else.

$$\triangle > \$$$

During the same years, our government officials and corporate partners juggled blame for the collapse of the middle class's greatest investment—homes. Houses are made of aqua-exports like copper, oil, and timber. When commodities increased in price, so did homes. After commodities surged, they sank, and so did homes. When houses plummeted, people blamed interest rates and the Fed, banks that were too big to fail, and predatory lenders, not water. Water was everywhere but not recognized. Homes and their interest rates became a usual suspect.

And moreover, considering how inflation was measured, was the CPI too low for too long, allowing interest rates to fall below normal levels? Did this invite homebuyers into the market who would not be allowed there under the current economic conditions? Did this falsely drive demand higher, and blow a bigger water bubble in the housing market? Ironically, the only way to alleviate the *market failure* in homes was to lower interest rates further—just another example of how government officials, *our economic refs*, buy time and juggle the pain.

Now that we see how inflation was marginalized *without* water, and how that impacted the CPI and affected interest rates, we can understand that this exacerbated the housing bubble. We blamed the

Fed and the mortgage companies, and never openly criticized the benchmark that determined those interest rates. Now *we can* probe the CPI and PCE about water.

Most people conceptualize the American dream as being married with children and white picket fences. Home values are the quintessential expression of the American dream and a core asset of every great nation. Because if home values sink, so does the entire wealth of the nation—it's scarier when we realize the stock market is also closely tied to real estate values. Once a home is bought, families elect to pay down their mortgage or invest more dollars into their home to make it better. Therefore, a home represents our biggest piggy bank, where so much of our wealth is held. Americans pour a large percentage of their after-tax income or potential savings into real estate in lieu of other investments like the stock market, which has a 5 to 8 percent annual expected return. And they do so without realizing how their homes fall in value against water and water inflation. Consequently, this North American regression continues as we embrace our twenty-first century. What happens when we realize our largest investment, our home, is being destroyed by water?

CHAPTER 11

House of Cards
Floating on a Raft

YOUR HOUSE IS FILLING with water. Imagine it's rising past your waistline to reach the bottom of the kitchen cabinets. The electricity is gone. It's darker inside the house than in the woods at night. Your children hang onto the arms of your spouse, not to play, but because they can't tread water. Hot showers are gone. Food is gone. A way to work is a joke.

When the water dissipates, everything in your ranch home is destroyed. You are losing wages at work, throwing away wedding albums, and trying to figure out who is going to pay for all of this. The mental anguish builds. You have to fight your insurance company every step of the way. Money is not coming back from FEMA to pay off the credit card balances charged to rehabilitate your house.

What happens if the insurance groups don't compensate you for everything you rightfully lost? You're exhausted and your body shakes with mental anguish as attorneys get involved.

The International Institute for Environment and Development
delivered a study years ago, stating that we are entirely vulnerable to
rising sea levels. In the US alone, this "Low Elevation Coastal Zone"
would displace twenty-three million people in areas such as New York,
North Carolina, and Florida. In 2017, we can add Texas to that list.

The fear that too much water stalks our home leads us to construct
paper dams in our minds to protect ourselves from storm surge, rising
waters, and saltwater intrusion into our fresh drinking supplies. In
polarizing fashion, home equity dives in other parts of the United States
where real estate falls off a cliff because of too little water. Drought-
stricken Georgia saw incredible declines in home values during the
2008 housing crisis. Other areas that come to mind are Arizona, parts
of California, and Nevada—homes built in a desert. They have very
little water of their own, with much imported from other states like
Colorado. The results of this entitlement:

> The Colorado, already taxed to the brink, is shrinking.
> Thanks to population growth, a decade-long drought
> and global climate change, the river is being drained

beyond sustainability. And if the Colorado were to die, Mr. Davis tells us, "it would be necessary to abandon most of southern California and Arizona, and much of Colorado, New Mexico, Nevada, Utah, and Wyoming."

Gerard Helferich, Wade Davis, *River Notes*

This scenario would collapse the value of many homes in America, yet we'd still be on the mortgage and on the hook for that debt. This is an example of water inflation eating away completely at our biggest piggy bank.

In the US, a fairly new hack at home values threatens those in many states experiencing the anger of Mother Nature. According to the American Association for Justice, "Allstate and its affiliates have stopped writing home insurance in Delaware, Connecticut, and California, as well as along the coasts of many states, including Maryland and Virginia. In Florida, Allstate dropped over four hundred thousand homeowners since 2004. The company ran afoul of regulators in New York for the same reason, and was forced to discontinue the practice." The cutting truth of climate change echoed elsewhere since 2007, where catastrophic damage, rising waters, and warmer temperatures threatened contracts of homeowners. Following the same tactic as Allstate, State Farm has embarked upon a campaign of insurance withdrawals and non-renewals in the aftermath of Hurricane Katrina.

Insurance companies and reinsurance companies are finding it difficult to calculate a safe level of premiums, since the historical record traditionally used to calculate insurance fees is no longer a guide to the future. For example, the number of major flood disasters worldwide has grown over the last several decades, increasing from 6 major floods in the late 1950s to 26 in the 1990s ... Insurers are convinced that with higher temperatures and more energy driving storm systems,

future losses will be even greater. They are concerned
about whether the industry can remain solvent under
this onslaught of growing damages. So, too, is Moody's
Investors Service, which has several times downgraded
the creditworthiness of some of the world's leading
reinsurance companies.

Lester Brown

It's crystal clear that rising costs and risks are *objective reasons*
to believe that *water inflation* is real, affects us all, and can even kill
corporations and their profitability.

If this is true, then shouldn't we also question the real value of
our homes? At the beginning of the twenty-first century, private and
publicly-traded insurance companies insured houses near or next to the
water's edge. Just years later, these corporations don't offer protection
in flood zones. That means profitable insurance companies won't dare
sell flood insurance. That's a red flag to us, because *if insurance giants
cannot find profits, how can we?*

Take for example some homes on the coast of North Carolina
without dunes to protect them. In this area of the United States, if
a hurricane destroys a "majority" of the home, the family is no longer
protected by FEMA because of its proximity to water, and can no
longer legally build on that land following a storm, according to Grist.
org. If storm surge *destroys the entire value of a home and makes that land
previously owned useless,* wouldn't you think a family would be devastated
by water inflation? This home and all the money put into it would
evaporate. This could be your family and their savings. This is what
happens when water inflation eats away at our biggest piggy bank.

$\lozenge{>}\$$

Imagine the housing market, all else held constant, as a victim of water's force. Picture one homeowner, in one of Virginia Beach's finest neighborhoods, known as Croatan. This area is pricey and shares some of the best box seats to the Atlantic Ocean that Virginia has to offer. In 2007, a man named Tom discovered that his homeowner's insurance, a policy through State Farm, was dropped after twenty-five years. Assume that Tom had never made a substantial claim on his property for flooding or storm damage. State Farm and other insurance companies no longer wanted exposure to the risk of water, catastrophic events, or climate change. Wind-driven storms that pose a threat to low-lying areas were creating losses for some insurers. According to Tom and the recent behavior of insurers, damages were outperforming premiums. Under these conditions, the flight of many insurance companies suggested unfavorable future forecasts for real estate near the coast. What does this mean for Tom and thousands like him?

At the time of interview, Tom began paying four times his old premium with a new insurance group. His premium *then* also increased accordingly, as storms over the Atlantic Ocean were named and upgraded to tropical storms and hurricane status. During active hurricane season, in a month when a category five is born, his premium is higher, even if this storm does not affect his property. Tom says that *luckily*, the federal government supplies flood insurance through FEMA to protect his home in the event of rising waters. Tom is a very smart individual, and neither he nor most people make long-term financial contracts supported by luck.

While that gives Tom peace of mind, FEMA is, in reality, a signal of a critical market failure, propping up home values where private insurance companies no longer dare participate, where profits no longer exist. FEMA is not only an angel where water is most likely to overwhelm us, but it's also neutralizing the effect of insurance giants skipping out on desert climates to avoid wildfire tensions (California). Because private insurance companies would be *insolvent if they protected homes they previously insured* (drought and flood-prone real estate), *the US government has to insure us.* In the end, water makes everyone the recipient of welfare, especially when saving a few homes saves the entire real estate market. If the government, whose deficits are already being questioned constantly, backs and insures the value of all homes for homeowners, then there is a very real and highly vulnerable asset of the economy on life support, subject to policy change and systematic risk linking so many dollars to uncontrollable forces. The US government is, in effect, propping up the housing market with debts taxpayers must absorb. This promotes moral hazard and makes housing a regressive investment for citizens.

Nonetheless, we will pay a growing insurance premium for all risks, as there are tiers to FEMA flood policies. This "program" of subsidization begs the question, is the house threatened by flooding and "superstorms" really worth the hassle or the risk? Unlike Tom, some people might not be able to afford a house with new astronomical insurance premiums and FEMA insurance to boot. What happens if the faith of the US government to support flood insurance is brought into question? How long will the US government continue to print money to pay for Mother Nature's climatic tantrums? Is the US government doing a good job at FEMA? And most importantly, how enticing is a home we love but can't insure? Or how amazing is a home that costs an arm and a leg to insure over the life of ownership? That reality eats away at all equity and gains, no matter how we slice it. It's a drain on our money and robbery of our wealth, a.k.a. water inflation.

There is no better example of systemic risk than the real house of cards that FEMA is subsidizing. If the housing market collapses, the value of the United States would be decimated. That is why economists lean on the *theory of the second best*. As Krugman and Obstfeld put it, the theory "states that a hands-off policy is desirable in any one market only if all other markets are working properly. If they are not, a government intervention that appears to distort incentives in one market may actually increase welfare by offsetting the consequences of market failures elsewhere." While they discussed the theory of the second best regarding free trade and tariffs domestically, it's time to take an already unconventional view of *welfare economics* and step outside the box with it, to show how the US government views *home economics*.

◊>$

In the case of the housing market, the government is applying the theory of the second best to the entire market, including what it calls "subsidized rates." And we are not just talking about the subsidy flood insurance that saved the insurance market from failing after Hurricane Katrina—we are now getting a little more specific, discussing homes built before 1978, when communities did not embrace a flood insurance rate map. Those homes are not in compliance with flood insurance rate maps for their respective area. They are much more subsidized than the general homes that needed to be rescued from the twenty-first century's flight to safety by insurance companies.

For example, a home in Lynnhaven Colony in Virginia Beach is a housing market where FEMA flood insurance policyholders are not paying rates that reflect their true risk. A fair example would be that a homeowner in Lynnhaven Colony might only pay $3,600 in annual FEMA premiums, even though that particular area required risk upwards of $20,000 to $30,000 in premiums. So not only does FEMA prop up the house of cards, it puts the reality of the housing market on a floating raft. The homeowner in Lynnhaven Colony does not get the real insurance bill, because if they did, there would be no demand for that home—essentially creating a run on the homes by the water, the same way people make a "run on the banks." Thus, *that type* of flood insurance by itself is subsidized to prop up the entire housing market.

In Virginia Beach alone, according to FEMA, the city had exposure in February 2017 of $6 billion for insurance policies and only collected $12 million in premiums. Meaning the city had only 0.2 percent of the cash reserves required for the exposure this coastal city faces as it awaits a hurricane or unexpected climatic event. Probabilities are rising! Whether it's Virginia Beach or FEMA, someone is on the hook for that difference. These insurance reserves are well below the Fed's "Reserve Requirements" for bank reserves, yet this is just *one coastal city* on FEMA's balance sheet. There are many, many more.

What about the actual payout when storms wreak havoc along an entire coast? Some American homeowners are still waiting on their insurance check following Hurricane Sandy and Hurricane Matthew. Others receive a small percentage of their home's value back from FEMA to rebuild. After Sandy, a home in Old Greenwich, Connecticut given a market value of $3,000,000 only saw $100,000 back from FEMA, but needed $1,300,000 to bring it up to present-day regulatory codes. Furthermore, the home was uninhabitable without repair and the electrical fires were real. When an individual pays $1,200,000 out of their own pocket to rebuild their home after one flood, and FEMA only guaranteed $100,000—that is proof of water inflation.

Present-day regulation from FEMA and water inflation are legit in Sussex County, Delaware, where Richard Heubeck shared his reality after Hurricane Sandy. "Legally, we're not supposed to be here," Heubeck said. "To lose all the equity you put into the house and be told you have to move out is something we don't even want to think about . . . We are really at the crossroads now."

And without flood protection through FEMA, if water ruins somebody's home, that somebody has no recourse for loss. Out of the many families who experienced the wrath of Hurricane Harvey in Texas, only 17 percent of such homeowners had FEMA flood insurance, according to the *Washington Post*. The majority, who lost their homes without insurance, will have to pay for the damages to rebuild, pray for government aid, or abandon their homes—and if the home wasn't entirely paid off, they still owe the bank.

All these examples and all these realities are water inflation at work without any solution. Perhaps the theory of the second best should be applied to something more important than measured risk in insurance markets and save more than just the housing market? Perhaps the theory of the second best should be applied to the water systems of the US (valuing water differently *first*), rather than react to the symptoms that this house of cards floating on a raft is going to get pummeled by bigger waves.

Hurricane Harvey, 2017

💧>$

Another relative opinion of the Hampton Roads area—which includes Virginia Beach and Norfolk—is that Virginia Beach obviously makes more housing "cents" for buyers. There are condos in Norfolk and Virginia Beach that fell prey to the soaring home prices related to very enticing interest rates and house demand rising. When the housing market collapsed, Norfolk home values plunged further than Virginia Beach home values. In fact, some home values today in Virginia Beach have climbed by 20, 30, 50, or even 100 percent, before the housing bubble, but not many Norfolk homes.

Some of Norfolk's homeowners are underwater and have begun reconciling their investment loss, as they outwardly announce it to friends and families. Norfolk is constantly flooding, sinking, and now considered a candidate for engineering marvels. Parts of Norfolk are below sea level and flood regularly. It can rain for fifteen minutes in Norfolk and you can walk outside to find your car fully flooded out—it happened to those I know many times. This is a sewer-system problem, and FEMA has high-risk insurance policies in the city. Norfolk needs underground reconstructive surgery to mitigate the rising tides. It needs to redirect the tidal waters that overflow city lines as water levels get higher over time. There are condos in downtown Norfolk that sold for $2,000,000 in cash in 2007 that go for less than $1,000,000 today. That's a 50 percent water haircut. There are homes that homeowners must rent out because if they sold them, they would admit a loss on paper.

After the housing market and stock market collapsed and then came back to higher highs, these Norfolk condos have not come back to original highs (value). It's theoretically possible that while Virginia Beach's houses are priced even higher, these Norfolk homes may never return to equivalent or higher highs because of water inflation—it's happened to homes in North Carolina. It's possible that water inflation completely deteriorates at the speed of their capital gains' timeline. Norfolk living is the quintessential meaning of water inflation as it relates to the asset value of homes. Norfolk real estate is subject to flash floods every time it rains and when the moon is full. Again, this is true water inflation at work.

$$\Diamond > \$$$

Virginia Beach looks like the smarter homeowner in the room relative to Norfolk; however, the value of Virginia Beach's house of cards is in many ways propped up and subsidized yet again by taxpayers. The tax dollars spent reclaiming the eroded beaches for homes sitting atop "prime real estate" facing the ocean are real money at work. Beach reclamation occurs when sand is pumped from the ocean floor back onto the beach, where it was naturally removed, and the beaches are thereby rebuilt. This is a fairly new cost, a tax we as citizens pay to finance uncontrollable evidence of water inflation brought on by Mother Nature. We ensure sufficient coastlines and prop up the value of homes and hotels owned by individuals whose speculative investment in land has now proved high-risk.

It is quite regressive to the lower income households if we consider how our taxes work in percentage terms and where the money from taxes are allocated predominantly. And we know it's convenient to justify the need to protect high-net-worth individuals to protect the jobs and home values of the lower and middle classes. Ironically, we are now aware that all American taxpayers are in trouble if the US government is forcing citizens to subsidize Mother Nature and the environment's unexpected claim to land. Moreover, how long can real estate hold value in terms of water inflation?

Just like the US Navy is raising its sea walls to prepare for rising tides, the concern of water inflation in Virginia Beach is alive. This area's deepest insecurities dance between too much and too little. The agreement for Lake Gaston's drinking water, shared with Virginia by North Carolina, robs Peter to pay Virginia Beach water, for now. As sea levels rise, saltwater intrudes into local aquifers, and low-lying regions like Norfolk lose more and more freshwater. When water for Virginia Beach is imported from Carolina and purchased from Norfolk, VB's waconomics is neither secure nor sound.

If the US government has to budget dollars the same way we have to as people, our housing market risks going on auction following foreclosure.

$$\Diamond > \$$$

What about human-made water inflation on the home? We call this result unrecognized leaks (ULs), leftovers from negative externalities (NEs). Thousands of miles to the west of Virginia Beach, in the glorious Rocky Mountains and on the Gold Coast, a similar phenomenon plays out. There are millions and millions of homeowners living among the Colorado River's influence and in California. In the twenty-first century, wildfires ravage homes in Southern California.

The Hollywood animated hit *Rango*, which starred Johnny Depp, covertly addressed the result of a water-poor Western US. This film *applied aesthetics* to pop culture's understanding of water value.

In *Split Estate*, Elizabeth and Steve Mobaldi left California in the early 1990s for the cleaner and greener pastures of Colorado. They were not privy to the oil and gas sector's effects when they purchased their place in Rifle.

It was Elizabeth who stayed at home while Steve worked. This exposed her to the home more frequently. In 1997 a water well was "blown out and contaminated" by Barrett Resources drilling the "BernKlau Well." Elizabeth lost her health and quality of life. The State of Colorado documented this situation. At first, the gas companies delivered drinking water to the Mobaldis, except that Elizabeth and Steve had to clean dishes and shower from that contaminated source. Dr. Daniel Teitelbaum, an occupational physician and medical toxicologist, explains, "There are a lot of sources of water vapor in the house; your dishwasher, every time you flush the toilet. You breathe it in and you absorb it through your skin. Your dose of the volatile organic compound from the shower water will be several times the dose you would have gotten from the drinking water."

The gas company claimed later that the water situation was repaired and safe to drink again. But the Mobaldis soon knew something was not kosher when they saw an oil slick on a glass of water they left out overnight. The following day they ignited a match to their water glass and saw it torch the air.

Elizabeth and Steve Mobaldi had to abandon their home, valued at $440,000, and moved elsewhere in Colorado. They were forced to walk away from their investment. This is an expression of the mental and physical anguish homeowners experience from oil and gas exploration—an anguish many clean-energy advocates deny, as it has not happened to them yet. Just like anyone in Virginia Beach may be insensitive to the reality of Flint, Michigan, for now.

After trying to sell homes following months or years of effort, foreclosure is likely for families like the Mobaldis. How many people can manage paying two mortgages, especially when one represents so much negative emotion? A family like the Mobaldis had to leave their home and pay one mortgage and one rent *just* to have clean water. We call this *hyper-water inflation*. Many people would not be able to hold onto their homes under such conditions and be forced to walk away. All the costs and mental anguish can be considered unrecognized leaks from the Mobaldis' wealth and health.

In the end, the message of many past and present homeowners subject to this UL ring in the ears of future homebuyers. Any prospective buyer should think twice about purchasing or renting a home for themselves in these oil and gas areas. Would the dangerous devaluation of our water supply not drive down the value of a home over time? After all, are the oil and gas wells going anywhere? No, they are multiplying, especially after the passing of the Energy Policy Act of 2005, as we'll come to understand later.

Now, think about the value of a home that offers no control over the future of its freshwater supply—a home vulnerable to droughts in California, or flooding everywhere. Maybe when the "housing bubble," or rather the water bubble, formed, it artificially (and dramatically) overpriced homes that suffer because of contamination from oil and gas, drought and shrinking water supplies from overuse, and flooding by water's force. Realistically, a home with that many skeletons in its closet should hold very little value. When its bubble pops, the house falls faster and further. At the heart of the matter, your home is becoming worth less, as too much water around you threatens its structural integrity. Inversely, real estate where the drinking supply is virtually toxic or nonexistent would hold no value either. Water is therefore quietly priced into the future value of our homes.

So, we must know how to alter water's force for our benefit, or too many of us will be financially punished. Let's find out how.

$$\Diamond > \$$$

We have learned that water was missing from the data that communicates inflation and therefore appropriate higher interest rates were not in effect at the dawn of the twenty-first century—rates that keep buyers who can't afford homes out of the market. If our measure of the CPI or PCE accurately accounted for water value and included a water quality index, interest rates would always have been much higher—meaning no housing bubble potential. Our *water gap* would be crystal clear to everyone if we increased interest rates to a normal level, because houses would be unaffordable to a majority of Americans under current economic conditions—at least, without a pay raise.

On grander scales, we *can* connect the impact of too much tidal water and too little freshwater to the value of homes. What happens if, over time, hidden water inflation has a higher percentage than the real return on your home's equity? Only falsely-deflated interest rates for home mortgages engineered by the Federal Reserve can offset this hidden tax on the American dream, and that's already being done.

For example, if hidden water inflation increased 15 percent in a year, and the net annual appreciation from a home, or net annual return on a rental property, is 3 percent, then your home's net return was a loss of 12 percent. This loss in equity assumes no new home repairs for that year, stable insurance costs, and taxes accounted for already. So that is a covert cost or loss that very few Americans are accounting for in retirement— *and no one realized it until after their home equity vanished in a flood.*

By now we are becoming water inflation experts. We get that a dollar today is not as powerful as a dollar in 1970—we prove this when we buy Skittles. So we know a dollar saved is eventually worth less later, and a portion of our shrinking buying power is an expression of water inflation. Even if a home value appreciated 300 percent since 1970, how many percentage points did the house *truly gain* with the US dollar's fall and the adjustment for an undisclosed water inflation?

People pay for water inflation and the water deficits that build from it. For example, a water deficit from Hurricane Harvey in Texas now exists. People take the financial blow for the team. Simply put, a citizen's goods and services are relatively more expensive year after year with water inflation—because water's costs are building. The value of that citizen's biggest investment and savings account, their home, gains less due to undocumented water inflation—such as rising insurance costs. Remember, as we print money, we are simply printing water to cover up this inflation hidden deep inside our wallets, perpetuating this financial drain. That's very damaging to the American dream, isn't it?

$$\bigcirc > \$$$

To challenge wa-conomic theory, I held an interview in 2009 with professor of economics James Roberts. The professor was asked whether it was possible that 2008's deep plunge in home values had more to do with water's force than financial bubbles brought on by predatory lending or the Fed's interest rate policy (based on CPI). Were home values in free fall *then*, in states such as California, Georgia, and Nevada—areas known for their increasing droughts? And what about in Florida, New Jersey, and North Carolina where fears existed that too much water would dislocate low elevated coastal cities? Did the housing market in 2008 severely punish the value of homes in areas limited by flood and drought, all else held constant?

In response to these questions, Professor Roberts said, "What would, could drive home values down is *rising* insurance rates on those properties. I don't think that's happened yet."

After I explained Tom's story to Professor Roberts, he agreed that these circumstances would decrease the demand for real estate and, all else remaining constant, lower the value of housing. Essentially, if the cost, or at least the perceived cost (and risk), of living there becomes greater, people would want to live there less often. And that would mean demand going down, and again all else held constant, that means price would go down. That was the theory then, and the consistent findings as of this edition of *Waconomics*. Norfolk's home values are just one example.

What this means is that either too much water threatening homes or too little available drinking water surrounding homes lowers the value of real estate over time. The lack of water supplies could be caused by contamination or pollution (Colorado oil and gas), or by greater, longer cycles such as drought (overuse in California), or by flooding (fill in your choice of coastal city).

$$\diamond > \$$$

Waconomics believes that we would *not* have to employ the theory of the second best for insurance markets alone if we valued water differently through the city-run channels to agriculture, industry, and residents (AIR). We would *not* have to be on the hook for so much water-inflated debt (such as the $6 billion in Virginia Beach) if there were a higher water benchmark for insurance agents to calculate premiums. Most importantly, if water rates were priced higher through the city, flooding brought on by sanitary sewer overflows from heavy rains and tidal water would be less probable (Norfolk), because our infrastructure and water systems would be upgraded and strengthened with the funds from the higher water rates to offset climate's demands (and destructiveness). We can apply the theory of the second best much sooner in the economic game of telephone, well before houses, and avoid this market failure in real estate altogether.

Because of water inflation, we know why the cost to insure a home is on the rise. We know why some homes currently are uninsurable. And we know why insurance companies can withdraw coverage from risky markets such as California, Florida, North Carolina, and Virginia, and have been caught doing so already. All else held constant, the premiums homeowners pay for private/public insurance are rising to levels which will decrease demand for that city's housing market. This will pull down the home equity value. Too much tidal water or too little potable water means that we cannot insure our homes *without paying a much higher price.* And that price is unknown. That rise in cost is water inflation.

If people cannot maintain the value of their homes because of external forces, whether those forces be climate or insurance costs, wouldn't that strengthen the view that water itself dictates more than the average global citizen admits?

What is interesting to note is that any time the average citizen has exposure to variable loans, those loans are tied to fluctuations in the US dollar. So when the mortgage industry suffered and the market collapsed in 2007 and 2008, fears sent the dollar higher and loans tied to interest rates ballooning. All of a sudden, houses were worth less, the payment worth more, and the only people around to absorb the loss were those on the mortgage. When they foreclosed and took a huge loss, the bailed-out banks got a hall pass. As in Chile, so here: a collapse in housing values equates to a collapse in our biggest piggy bank.

Some will argue that water is *not a public good*. We should ask those individuals if real estate is. If we've propped up and saved real estate through government's FEMA, and homes are infinitely penalized by too much water and too little water, then we should just save the water systems instead, and the rest will follow downstream.

CHAPTER 12

Why Fix the Pipes?

DO WE KNOW WHAT lives inside the walls of this country's aging pipes? Do we know what these pipes pass along to us? Do we know how much water they *leak* a day? Think about hospitals above the ground. Consider the diseases actually given to patients in hospitals, which are cleaned routinely every day. Consider the out-of-control costs that *leak* and spiral unto the bills of those patients and insurance groups. Pipes are no different, but unlike hospital visits, our susceptibility to water is constant. The pipes, I believe, are the arteries and veins of the hearts of our families. It's time to clean up our body's transportation highway.

Like a regressive tax, the income-to-water rate between rich countries and poor countries demonstrates how the poor subsidize the rich. Highly water-intensive goods produced in developing nations flow downstream for consumption to highly developed countries at deflated prices, below US environmental standards, and without compensation to the people of that poorer country. Corporations globally integrate and relocate to underdeveloped nations, to export their need for cheap, unregulated freshwater for production means. Waterinfo.org has reported that "90 percent of wastewater in developing countries is discharged into rivers and streams without any treatment," because those nations lack sufficient environmental regulations. Water.org connected the results: "Half of the hospital beds in the world are occupied by patients suffering from diseases associated with lack of access to safe drinking water, inadequate sanitation and poor hygiene . . . More people have a mobile phone than a toilet."

Many corporations are getting their way. We can, too.

Would the United States be so productive if its citizens had nowhere to put their feces? And if they had to take someone else's water and aqua-exports? That's what makes the United States of America a great nation—our power to offer the greatest infrastructure in the world at a great price with options. Today, "somewhat" safe drinking water and necessary sanitation are pillars of great nations. Unbeknownst to most, drinking water and sanitation have been the most vital factors to the economic success of US history. Yet the water infrastructure investment for citizens is falling to corporate privatization, just like the bridges and roads. We are getting sold out.

A huge portion of the overwhelming debts of US cities and states are indirectly connected to water inflation and US water deficits. For example, the people of California have physical water deficits from overdrafting their aquifers. The non-physical consequence from their water deficits shows up in economic underdevelopment, as California lacks sufficient dollars for public schools, which indirectly results in a higher crime rate. Regardless of how deficits are literally or figuratively created and connected, too many IOUs from the city or state only give the privatization model momentum—a cause to justify public change.

The more water inflation, the bigger the water gap, and the bigger the deficit. As the world's water deficits build, there will be corporations waiting for cities with open briefcases of money, ready to be their savior. We could parallel the results of these long-term deficits with the lease or sale of timeless assets such as US bridges and roads. These were, at one time, basic public duties. The US Government was born to create an environment for the private sector, offering water and sanitation, airports, roads, connective bridges and tunnels, and deep ports which allow capitalism to thrive continuously. That's what is meant when they say "less government." Give the citizens the infrastructure and let them innovate and inspire the globe. That's the American dream in the end.

How the hell did these tangible public duties instead turn into the intangible roles of allocating digital dollars for entitlements, tax breaks, and subsidies? Of late, it seems like we do more of the latter. So why did the US relinquish its public duties? I believe it was because water value was born in the 1970s and the US started suffering from costly regulation. And I believe it is our civic duty to connect our value and worth to water's value and pipe a connection that will reverse the trajectory of history.

Would you rather pay a toll to your country when it would return the benefit to you later and your children, or a foreign corporation which will solely benefit select shareholders? When we privatize our pipes, there is no guarantee that non-revenue water, a.k.a. leaky pipes, will ever stop, but there is a guarantee for a shareholder leak. Shareholders will receive some fixed percent of water profits. If the leaky pipes are not fixed, that means that 10 to 13 percent (shareholder leak) *plus* a rounded 30 percent of our water could leak from our wallets and no one but attorneys will get a say about the price hikes to come. We'll see that dynamic later in Rockland County, New York, barely north of Manhattan.

$$\Diamond > \$$$

Let's speed things up as we imagine a country where only the finest automobiles are driven. A nation comprised of high-speed cars which drive low to the ground to enable handling and performance. These models on the roads are Audis, Porsches, and Ferraris. Only these fast automobiles are driven because this country believes in efficient time on the road and great individual results when driving. And such required performance must be maintained.

Now picture this country with a growing epidemic—it's becoming a pothole nation. Each street and every timesaving highway are littered with a dangerous number of potholes. One of the sole responsibilities of government and all governing bodies is to provide a business environment that allows for more seamless performance and growth. That performance rewards each car and inspires innovation of a faster more efficient country.

So, let's pretend that we are the Ferraris of this nation and that those potholes are the water pipes. If our bodies were these Ferraris, then wouldn't we demand our country not only fix the pipes but consistently maintain the pipes so that we could keep performing well? Shouldn't our personal performance matter to us? Wouldn't your performance matter to your husband or wife, son or daughter, grandchild, or great-grandchild? Do we want life to improve, if not for our own welfare, then for our families' future? We want to be alive to hold, know, and love our children's children—to be great-grandparents. We want to be remembered and for our stories to carry history. That is wealth. That is everyone's legacy: the legacy of appreciation, joy, and love—more of the dream.

Now let's assume the US is a population of human bodies, each best supported at the core, in its abdomen, with many individual and powerful muscle fibers stabilizing its structure. Like a human body, the US is operating at its optimum level when the core is educated, prepared, and trained to work. Let's call that the *middle class*. If the middle class shrinks, the lower classes will increase. Symptoms of such regression include food stamps, increased crime, and the rich getting richer without a relative gain for the people. But if the middle class rises, the lower classes will decrease as they are pulled up into new income brackets. Manifestations of such progress include affordable food and the ability to vacation. The core of the nation can make the lives of people around them better, as true citizens should.

The United States of America, like every great nation, shares a common thread in conjunction with its relentless prosperity: its middle class. The middle class has been and will always be the fiber of great nations who lead and of empires that last. Without a middle class, no country is a true world leader. A great nation must show its people that it believes in them by investing in them. The birthright of the middle class exists because of the fundamental public duty of governments, their bread and butter of tasks, to provide infrastructure. If the middle class is the abdomen of every great nation, then infrastructure is the spine and the pipes are the veins which pump blood to its heart. If those arteries and veins are clogged, then we will die much sooner, just like other empires that fell.

Without safe water and quality disposal of wastewater, the United States would be too busy fighting disease, becoming an unstable nation at the core. During the 1850s in Great Britain, Dr. John Snow realized this about water and cholera. Cholera was taking lives through poor sanitation and unsafe water. The results were as deadly and destructive as wars. Pandemics undermine stability, and not until Great Britain and the United States improved their water systems for clean water and sanitation did these nations offer an environment to safely thrive. *And the 1850s are also the last time that many pipes in the United States were touched and upgraded!*

Today, the greatest world war is ongoing in underdeveloped nations between people and water. Safe distribution of water and proper disposal and treatment of wastewater limit new and old, acute and long-term diseases—economic imbalances. For hundreds of years, that's what great nations have been fighting for. They have been containing and fighting back death and disease. Today, waterborne diseases are still the deadliest threat in underdeveloped countries. So now we can agree that all classes in all countries, developed or developing, are less powerful in economic and health terms without some level of water and wastewater treatment.

The US would never have been a great nation without the

infrastructure for drinking water and sanitation; diseases and deaths would have kept economic handcuffs on it, like those that underdeveloped nations face today. Developed nations depend on this security to progress. Older empires knew this. In fact, there was no better platform to springboard into greatness than the pipes underground, like the rivers that predetermined the lands of empires and kingdoms. Many historical inventions were important, of course, but without access to clean water and sanitation, we could not optimize any of those inventions. Without the pipes, our electricity, healthcare, and housing development would all be less potent talking points.

$\lozenge > \$$

The US is a top-heavy water consumer, prospering as it leans on other countries for cheap water. As developing or poorer nations adopt developed nations' contagious water-intensive habits, water is going to squeeze everyone. This diet is ferociously nurtured through marketing directives by MNCs (multinational corporations).

The covert requirement to increase the price of water is upon us, but how we develop the price of water will create one of two economies:

1. Private control and irrational movements in aqua-exports, further advancing corporate feudalism

2. Higher wages in developed economies due to proper and transparent water pricing and potentially healthier developing countries

Water and wastewater are one of the few industries that absolutely require government control. Waconomics advocates that water rates must rise in the US through municipalities run by cities. This finally puts a surplus in the water and wastewater coffers for infrastructure and operations, a real current-cy for economic development.

Remember, "on average, every US dollar invested in water and sanitation provides an economic return of eight US dollars"—an 800 percent return on our water rates. This is why the price of water can rise through the public sector in the US and help Americans, rather than the hurt everyone expects and points to before actually raising rates. Over the years, our mobile devices for communication and their data bills—which did not exist for much of history—have risen exponentially and we don't even flinch at the costs. Yet everything is not okay when the water bill goes up, right? We'll call the municipalities, and when out with friends, we'll have sidebar conversations about our increased water bills.

Maybe we would like to know what we are getting back for a higher water bill? We can rationally explicate that by investing in water and wastewater and protecting this service in a local government haven. Such loyalty for recognizing water value would create true buying power through increased real and nominal wages; we'd get a raise, and much, much more.

"The Environmental Protection Agency (EPA) estimates that water is the third single largest expenditure in our entire economy. That means that it falls just behind defense and Social Security," wrote Kevin Kerr. Yet a UNESCO 2003 report found the United States ranked near the bottom of developed countries for water prices.

Higher water rates allow and generate effective and quality infrastructure, and will *not* induce economic pain in the long run. Contrary to sensationalized opinion, higher water rates will invite exponential growth and less hidden water inflation—"Made in the USA" could truly rise with a return to efficiency. Citizens of every city should plead for their municipality to continuously run their water and wastewater system. Privatization would cause them a much greater cost in the end. Just imagine flashing water prices like gas station lights.

Today, it could be explained that developed nations regress in family size (fertility rates), wealth, and purchasing power because of water inflation. The impacts of water inflation cause unrecognized water value to rise in the prices we pay for goods and services like healthcare and Skittles, matched and multiplied by all countries catching up to developed nations' status with their mirrored demands and their new regulations. Everything is going to get expensive because of water's cost, whether our government tells us this or not. Based on current water demands, emerging markets will see an escalation of their middle-class numbers. But they will most likely not propel forward with the same luxury and speed of previously developed nations. Yet it's possible that all this can change. Waconomics suggests that solutions are a derivation from the pipes.

◊>$

The United States of America and its legacy were interwoven by great risks taken. *Safe* was not the roadmap for the American dream. Safe is a compromise, a concession, a dance with fear. The US seeks the best innovation day in and day out. That's the American dream. So do you believe that we should play it safe with our pipes, or take a risk for our families and our legacies?

We have just one water future with very few choices. We can let the water and wastewater industry run our lives and let life become more expensive, financially regressive, and dangerous to our health. Or we can decide our future and how we prosper. It is said that you either create your dreams or let someone else decide them for you. Well, that decision is here. The price of water will rise, but on whose terms? For your water and sewage needs, would you rather invest in your country and your family's future or invest in a strategic business unit of a multinational corporation from France? Would you want to lead your family with awareness or be led blind? Are we the decider of our dreams or a pothole nation?

If you want less government and more return from your hard work, we must ask for water rates to rise now, but under government control. It is possible to give the government a little more money up front and get much more back from our more profitable industries. Remember, $1 now for $8 later. Unfortunately, the more we allow the US government to determine our families' needs based on their statistics on inflation without our input, the greater the need to privatize all aspects of our families' lives—including water.

Water prices' rise is inevitable. We cannot choose not to allow it; we can only choose how to allow it. How we make that choice is the reason I am writing you. Water does and has created financial change to our lives and the lifetimes of our families. I believe we can give every class and corporation a raise through water, and it is here that we will evaluate why there is a choice, how that choice can be made, and what that choice means.

CHAPTER 13

I Went Undercover

I WASN'T THERE TO make money. Don't get me wrong, the pay was a perk. But I went undercover for a story. This journey wasn't for me; it was for something greater than me. I knew that water could save the American dream. Usually, we are afraid of being wrong, but my fear was that most wa-conomic theories would be proven *true*. If they were, then I would be obligated to finish this book—at all costs.

If I wanted to prove a water gap in our lives and how the water systems predetermine our wealth and health, I would have to get inside the public water sector—which is precisely what I did.

The water districts, a.k.a. municipalities, are such a key player in lifting up families that I had to put myself in municipal shoes. It was time to see for myself the economics, finances, infrastructure, and, most importantly, politics from a local water district's cubicle. I wanted to help the world see that water can be used as a tool to rebalance all economies. So I continued to pursue my American dream— to explore my visions of an economy thriving on water rather than surviving with unsolved mysteries. To help in this, I created a desk at the accounting and finance division of the Hampton Roads Sanitation

District (HRSD). At times, it's appropriate to follow the money, but for this mission, it was more appropriate to follow where the money was counted.

So in 2010, I called HRSD and planted the seeds for this story. At the time, no job opening or internship existed in the accounting and finance division. Yet after a few conversations with the chief financial officer (CFO), I had an interview, then a lunch, and finally a project-oriented position just for me while attending Old Dominion. I worked directly for the CFO, and I reported only to him. I often asked a lot of silly questions. He jokingly called me the "intern." I did play the intern role perfectly. In hindsight, it was my first appearance as an actor.

◊>$

What exactly does HRSD do? It treats the stuff that goes down the drain. It manages the flooding streets. HRSD addresses the water overflow from storms, a.k.a. sanitary sewer overflows. It manages the water that flows back into the ecosystems of natural environments and urban living. HRSD is unlike many sanitation districts, in that much of the total overflow from Virginia becomes the burden and responsibility of HRSD as a result of the Chesapeake Bay Watershed. If too much chicken poop or pigs' blood leaches from factory farming in Virginia, it is a valid concern for HRSD. If too many herbicides or pesticides are utilized on foods for farming, that is a valid concern for HRSD as well. Most of the waters from Virginia—above ground and beneath the surface—flow into a runoff for which HRSD is responsible: the Chesapeake Bay, one of the deepest natural ports in the world.

This is a big job. and the EPA holds HRSD highly accountable for such roles. That is why HRSD's wastewater rates are some of the most expensive of all sanitation districts in the country. It handles the negative externalities (NEs) of all the farms, industry, and households for the entire southeastern state and then some. Therefore, this region pays up for the spillover benefits (positive externalities), such as a

clean bay to swim in. HRSD manages wastewater treatment services
for seventeen counties and independent cities, and various military
installations throughout southeastern Virginia. Virginia Beach is
very similar to California's San Diego, both in military presence and
watershed responsibility.

At that time, I was hired for a specific project regarding the
efficiencies of consolidating all cities' operations under the HRSD
umbrella. Once my task was complete, the job was done. While the
investigation was ongoing, as I simplified my theories and worked
religiously on waconomics, the results were made public before the
printing of this book. The purpose of this project was to demonstrate
economies of scale for the ratepayer.

My job requirements were to collect and review the cash flows
and income statements of all these seventeen cities and counties for
a five-year period. I dissected data on water and wastewater, cost, and
revenues, and extrapolated information from each locale's twenty-
year capital improvement plans (CIPs). These CIPs are based on very
old pipes and new community developments. CIPs expand with new
designated infrastructure demands, mandates, or repairs requested by
the EPA. Sometimes these improvements become prescribed through
consent decrees—a civil case or settlement with the EPA. The decrees
are proposed to improve or replace the infrastructure of the pipes
under each city and county. There is always a progressive standard and
a constant upgrade required. This is due to the expiration of aging
pipes in the US and new developments herein.

But what I learned at HRSD was exactly the information I was
hoping would be *untrue*. Unfortunately, HRSD validated what I
believed was quietly happening in general to water rates and the water
and wastewater industry as a whole. So let's review my *expectations and
confirmations* from this experience, and then the *results* of playing the
"intern."

$$\Diamond > \$$$

We pay for the water we use, and therefore the water HRSD treats. Historically, HRSD and other water districts gave corporate-discounted water and wastewater rates because companies would agree to use large volumes of water, which guaranteed revenue for HRSD—essentially establishing a price for bulk purchases. It's justified the same way paper towels are priced at a wholesale club versus a 7-Eleven. Imagine corporations paying wholesale water prices and residents paying 7-Eleven water prices. *Idealistically speaking, the discount rate incentivizes overuse and a cost gap, especially because water doesn't go bad and it's not going anywhere.* When a company offers long-term employment to a city— for reasons disclosed below—cities would make concessions for their operation with discounted water rates (or tax breaks). Each company might negotiate, or already have a separate water and wastewater rate than residents. This is based on past practices that have gotten us to this point today.

This history admits that private businesses expect *water stamps* just as much as the people waiting in line for *food stamps.* We're all the same, right? Cities and states in the US have been overly accommodating to business. Perhaps this is a result of our national *insecurities* related to US environmental regulations? Remember, the EPA gave water a higher value. So corporations cannot afford to make a profit without a difference in how they are charged water (or taxed) from how citizens are charged water (or taxed). Corporations cannot afford production without a subsidy first, to make up that cost difference or cost gap from the 1970s.

This is the beginning of the *water gap* and "stamps," where both corporations get welfare and then people get welfare, but at different times—Republicans and Democrats seesaw. Corporations, started by individuals, lobby and negotiate for discounts and tax breaks. For example, steelmaker Alcoa, carmaker Ford, and computer maker Intel get billions in subsidies. Later, that dependency trickles down the corporate pyramid to the people at the base who need entitlements and handouts. It's a culture that is passed down from the top. Part of the discrepancy is that peoples' salaries, unbeknownst to them, are not adjusted for water inflation. For example, Skittles increase in price at a faster rate than the minimum wage. So the people believe the goods and services they pay for are "expensive." Regarding our water gap, this is the truth; because water went missing from our wallets, the prices of everyday goods and the cost to produce them had to be subsidized for corporations through stamps and tax breaks, basically entitled to antiquated costs, that made production artificially too cheap today. If corporations did not have a subsidy first, then goods and services would be even higher in price than they are. That's how we know inflation readings for decades were inaccurate; otherwise, corporations would have been allowed to charge higher prices and *would not* have to negotiate with the government first for discounts, intervention, tax breaks, and water stamps.

It's important to see negotiated water rates like tax breaks that benefit corporations. If you're unaware of this, simply check out New York's tax breaks or Delaware's corporate tax haven. States and nations are all competing for businesses, causing society to think corporations need a handout to be successful. How is there a water gap in our wallets? That can be explained when citizens are expected to pay full income tax, but some corporations are not. These corporations are thriving and a majority of citizens are hurting. Citizens and their families buy goods and produce things just like a corporate body—households are tiny little corporations, too. The bottom line is that residents of a city have paid a much higher water rate, meaning those with the least individual purchasing power pay the most—like grocery shopping for your family at 7-Eleven. Basic economic principles, right? The big users pay the least, which means poorer individuals are subsidizing bigger, richer users. We realize that all lives are created equal, and with water, pricing should match up.

Ironically, this competition between states and nations grows for new businesses. While we hear blanket statements about the US government's entitlement and subsidy programs, we hear less about why exactly corporations accept discounts and tax breaks. For example, according to Philip Mattera at Good Jobs First, 75 percent of *total* disclosed subsidy dollars, equal to $110 billion, has gone to just 965 large corporations. The Fortune 500 are included in that number. In the private sector, corporations would not produce goods if the costs to produce were too high. So production itself is unaffordable without subsidies. Welfare simply begins with the corporate relationship. Again, the corporate commies are getting in line. Why? Because water inflation is affecting everyone.

We could end much of this welfare through waconomics. The solution addresses citizens and corporations alike. When we disclose how, we'll certainly reduce the need for water stamps and food stamps.

$$\Diamond > \$$$

What else did I expect coming into my "internship?"

Cities, on average, have pipes in the ground that far exceed their life expectancy. Pipes that bring water to and wastewater from homes were ideally only meant to last fifty or so years. Generally, pipes in the US exceed that lifetime at minimum by another fifty years, with many of those circumstances in major cities. "Some of America's municipal water systems are so old they were built during the *Civil War*. Result: a significant water line bursts on average every two minutes somewhere in the country. In some places, like Washington D.C., a major pipe breaks every day," according to Sean Brodrick. For the nation's capital, the winter weather of 2014 exposed such infrastructure truths. "Nearly 600 water mains of the Washington Suburban Sanitary Commission (WSSC) broke or leaked that January," according to *Managing Climate Change*. "In December 2010, a record 645 breaks and leaks occurred. The extreme rise and fall of temperatures placed stress on the pipe walls, causing them to expand and contract."

An important first step in addressing the problem is looking around the country at the risks water utilities

are facing in terms of how the physical infrastructure—
water mains, sewage pipes, storm drains and treatment
plants—are faring under severe weather. For instance,
are they failing under the effects of extreme heat? Are
they able to manage the massive amounts of excess
water from flash flooding and prolonged rain storms?
Western states such as California, Nevada and Texas
have been plagued by severe drought. In California,
this has led to water rationing that is causing main
blowouts due to water pressure buildup that occurs on
days when water bans are in place. In Texas, extremely
hot temperatures are causing more pipes to distress and
break.

Water Finance and Management

All of this confirms that pipes in the US leak anywhere between *10
and 50 percent* of the water moving underground every day. We know
this to be *non-revenue water*. Before coming on board HRSD, I looked
at non-revenue water as inflation, especially the water that is leaked and
lost from fiscal inefficiency and oversight. After all, societies around
the world have experienced inflation and hyperinflation because of
government fiscal planning and policy, and a city municipality is no
different. Global non-revenue water estimates reach upwards of 30 to
65 percent. As *Water Utility Infrastructure Management* reported:

> In other words, 34 percent of the water that a utility
> has treated and pumped is not billed. The World
> Bank estimates that non-revenue water costs utilities
> $14 billion, annually, worldwide. In the United States,
> reports have shown that the non-revenue water totals
> between 10 and 30 percent, with more than 240,000
> main breaks each year. Many utilities are dealing with

pipes that are more than 100 years old and in desperate
need of replacement. These repairs are expected to have
up to a trillion-dollar impact to utilities over the next
25 years.

So that's the general foundation of what I expected to confirm at
HRSD. We'll get to the theories after we talk about some results.

◌>$

To my surprise, HRSD was not incentivizing overuse. HRSD
no longer discounted sewer rates based upon usage. In Williamsburg,
beermaker Anheuser-Busch had previously used one hundred cubic
feet of water (CCF) at a negotiated rate. That was no longer possible
under the new HRSD regime. This was a relief for me. My admiration
for HRSD grew.

So, why would municipalities want to discount water rates to
attract new companies that offered long-time employment? If a water
district is the recipient of a new enterprise that creates jobs and stable
employment, then this invites longtime residents, which generates a
strong demand for the housing market. These homes prompt a need for
consistent water use, and thus, a longtime employer makes strong water
users. Longtime water users equal consistent strong revenues, strong
cash flows, and therefore a healthy municipality. The stronger the cash
flows, the higher bond ratings a municipality receives.

Improving and replacing existing pipes is a mounting responsibility
that local governments must embrace. Maintaining the physical integrity
and efficiencies of the pipes creates a stronger balance sheet. The stronger
the balance sheet, the higher the bond ratings a municipality receives
as well. Bond rating agencies shape investor confidence, and those
investors help fund all operations and maintenance. A municipality's
financial performance, audited yearly, will determine their credit ratings
as delivered by groups like Standard and Poor's. S&P evaluates the

risks associated with investing in municipal bonds for specific regions of the US. The higher the ratings, the lower the risk for investors, and the cheaper it becomes for a water district to take loans out from investors—and so they sell bonds.

Efforts for operations and maintenance are funded two ways. A municipality's budget balances revenue from the ratepayer (water rate), with the additional responsibilities of selling bonds and paying interest to bondholders (bonds). The requirements for capital improvement plans (CIPs) have costs which exceed the water revenues brought in from all the ratepayers. Our water rates that are printed on utility bills are expressing new and old debts created per city, per year, to fund operations and build new infrastructures. They include interest obligations. And tax-free municipal bonds are necessary to fill the hole of financing to meet the EPA mandates for twenty-year capital improvement plans. Corporations (agriculture and industry) and people (residents) are the ratepayers (AIR). And we are generally not paying enough to develop new communities, maintain the pipes from existing communities, and replace the old ones (especially when we factor in discounts).

Simply put, there's a water rate issue we must address. And bond sales become the life raft of the operations, propping up the shortfall from the water rate coming in. Bonds are compulsory to raise enough capital to fund operations and install EPA-mandated infrastructure requirements and updates.

For the water and wastewater sector, there are two bonds that strike me as important: general-obligation bonds (GO bonds) and revenue bonds. These financial instruments are known as municipal bonds, sewer bonds, and the like. Typically, investors are guaranteed tax-free interest on some of these bonds and principal repayment at the end of its maturity. This is a complex matter, yet an obvious means of financing operations with more debt.

Consider interest discussed here as a form of inflation upon our families—like an unrecognized leak. *No one asked us if we minded paying a compounded or multiplied inflation, like the compounded interest from the financing of our pipes.* Oddly, most financial advisors will tell you that to plan for financial independence and retirement, you must eliminate as much exposure to interest (debt) and taxes. As noted, corporations already have the advantage with water stamps and tax loopholes. So why doesn't the city give us a heads up and ask whether we'd rather pay this much upfront in a month for our water rate (without interest), instead of this ridiculous lump sum of billions in interest later in receiving a lower water rate today?

If a municipality or private-public partnership such as HRSD has a stronger balance sheet, it gets good S&P ratings and will receive the attention of the right investors—raising enough bond sales to complete the necessary updates. Fortunately, the seven cities of Hampton Roads pay lower interest, because they are populated by all branches of the military, farming, government contractors, higher education, ports and shipping, and tourism, which drive jobs, housing, and life-long residents to the community—water users. These conditions stir an appetite for investors and a safe bond market. Yet most US counties and cities are not as pretty on paper as Hampton Roads, including Detroit, industrial cities in the South, old coal towns in West Virginia, and steel producers in Pennsylvania. If the city or county is facing economic hardship, loans are more expensive—meaning interest rates are higher, and people get taxed more in the long term on that compounding interest. Water inflation builds and builds upon this history. Areas like these are no longer as strong in industry since the 1970s because of the EPA, and employment and housing have thus weakened. The municipalities in these towns are at a disadvantage. For these cities, it becomes even more difficult to justify higher bond ratings from rating agencies like S&P, and thus they just get higher interest rates with "jumbo" loans.

◊>$

My fear at HRSD in 2010 was that most of my theories would in fact be validated. If so, then I would be obligated to finish this book for the people of the United States. While my ego was afraid of being wrong, I wanted to be wrong more than to confirm that my theories were the truth.

I verified that there was an overwhelming political force holding rates below what was necessary to finance the operations of the sanitation district. This was one of the primary reasons I was there and one of the stories I wanted. The essence of this political force was of most interest to me. Yes, raising taxes is difficult and many times capped, but raising water rates is a measure of value. And value must equal price and time. The longer pipes have been underground without updates, the higher the value or replacement price should be. Additionally, new community development adds more value to water rates. Recognizing water value calculates all that time in the ground and doesn't discriminate against reality if required replacements or new costs appear high.

From my discussion of the rates with a co-worker, politics emerged. The more I listened, the more information flowed. I inserted lighthearted, watercooler curiosity and fed a nonchalant conversation. When it happened, at first, I couldn't believe it. But it checked a major box for wa-conomic theory.

There were political officials willing to stand on their soapboxes and promote sufficient water rate increases to AIR. These politicians were people just like us, who believed in sacrificing incremental dollars and cents for our families and our future generations. But these political efforts were smeared. These political figures were either coerced into less passionate platforms, threatened with the loss of their office, or on one occasion, actually removed from their political seat. This occurred in the Hampton Roads area. Why would political passion to raise water rates be smeared? Who told me this? Were they a credible source? Of course they were. But I am not naming names. And furthermore, it would do little good, because this book will never change that aspect of political will.

The more important question is, why would rates be intentionally held down? The importance of paying for the upgrades of this nation's infrastructure should clearly take center stage. Remember, it's about us. It's about our dreams. Certainly, aging pipes, the EPA, and capital improvement plan requirements mean that water and wastewater rates need to rise. Yet the influence to maintain wastewater rates below HRSD's breakeven revenue was strong. Rates were being suppressed when they clearly needed to be raised so that infrastructure improvements could be funded. I wondered how HRSD could eventually replace all the capital improvement plan under their consent decree and build new infrastructure for growing cities, especially with an intentionally low rate to users. And as I mentioned before, HRSD is one of the most expensive water districts because of the Chesapeake Bay Watershed. It is a leader. Could this be why political forces cap rates in Hampton Roads? Unfortunately, this is a tension point for not only HRSD, but all water districts around the country. HRSD already had relatively high user rates. But what about old car, chemical, and coal towns? How would the Detroits and Pittsburghs raise enough money for water and wastewater if their cities were failing after the industry resigned and their jobs exited? Especially if those industries depended on water stamps and tax breaks, only to leave behind a toxic graveyard? Did these cities get upside down with their finances because water rates weren't high enough to fund a water system to handle corporate participation before 1970 and the next generation of families and businesses to follow?

How can a town or city overcome corporations leaving their communities because of new environmental regulations, and then later deal with the EPA's requirements to address the mess? Quite the Chinese finger trap. Even more ironic is the burden communities still feel today because of the industrial pollution that remains from that corporate participation. Because water remembers.

◊>$

HRSD's rates were constantly being revised upwards with yearly increases. But were wastewater rates matching non-revenue water gaps like those 10 to 30 percent leaks? On top of that, were wastewater rates tracking the same aggressive, inflationary pressures such as *headline* inflation? No; at HRSD, they were not. Rates were measured with *core* industry inflation estimates. We now realize core inflation benchmarks lack water and under-represent our true exposure to water—remember "volatility." Because HRSD's wastewater rates did not track headline inflation, there was a water gap in the calculations themselves.

At HRSD, I kept on probing the CFO with fun "intern" questions. Why were the rates not using the most aggressive inflation benchmarks like the headline inflation—after all, aren't food and energy prices primarily made of water? Suggesting new economic theory like water inflation was outside my pay grade, and convincing a director of a major municipality even more challenging. But challenges are made to be overcome.

INFLATION DIFFERENCES
a.k.a. Price Change

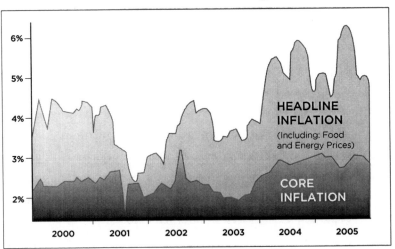

So from this view, the difference in the "headline" inflation and the "core" inflation is water inflation, and a large portion of the water gap—and non-revenue water (NRW) and ULs (unrecognized leaks) could speak for the rest of water inflation.

There's a way to fill the *water gap* created by water inflation: address the bigger elephant in the room, *a suppressed water and wastewater rate*. If we recognize water value through the water rates, then we begin to include all components of water inflation, like NRW and ULs, which have been MIA. We can end water deficits—filling in the cheese holes of the US economy, ultimately giving cities, corporations, and ourselves a raise and simultaneously reducing welfare: a win-win-win.

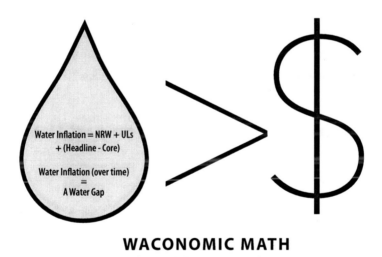

Water Inflation = NRW + ULs
+ (Headline - Core)

Water Inflation (over time)
=
A Water Gap

WACONOMIC MATH

What I will suggest in the short term is wa-conomic math. This simple addition is entirely conceptual and not absolute for any water district. An existing water or wastewater rate should include the difference between headline inflation and core inflation readings (from the BLS or BEA) on top of a variable water inflation (that includes losses caused by non-revenue water). For example, 1.5 percent (headline CPI—core CPI) + 4 percent (water inflation) = 5.5 percent (the conceptual increase in one district's current water or wastewater rates).

Eventually, we must get inside both the CPI and PCE and create a water quality index, which will address so many ULs in our bodies and our economy.

$$\mathbb{0}>\$$$

Haven't university costs exponentially increased since the 1970s? Aren't hospital and hotel beds more and more expensive each year we use them? Education, healthcare, and hotel inflation assume huge increases over time that people notice and recognize. Yet families still pay to educate themselves and their children. And people still pay to vacation with their families. As they should. Please notice the timing of when college costs took off, according to the College Board.

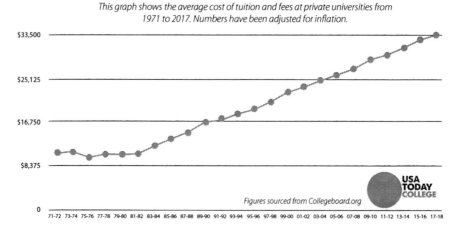

Average Cost of Tuition and Fees

This graph shows the average cost of tuition and fees at private universities from 1971 to 2017. Numbers have been adjusted for inflation.

Figures sourced from Collegeboard.org

USA TODAY COLLEGE

Does anyone know why healthcare, hotels, and universities *can* express inflationary reality with their sticker prices, but industries like water and wastewater, which make those other industries possible, *cannot?* Well, hospitals and hotels must remain profitable to stay in business, and education constantly requires more donations and subsidies. Moreover, many education facilities are operating as nonprofits and given a tax break. Even with their tax-exempt status, their costs and prices are exponentially rising. Why can't water rates? If our financial advisors *must* beat inflation so that we can successfully retire, shouldn't our pipes beat inflation, too? If they don't, then even our children will be left behind. And cancer is proof.

While many short-term diseases like cholera are now a thing of the past, other diseases seem to be ballooning. A perfect example of such is mycobacterium avium complex, known as MAC. It's a very new disease related to water and actually prevalent in hospital water. While it is treatable with lots of antibiotics, there are doctors today dedicated to just this disease. Talk about investments in time and specialization! The cost of healthcare and health insurance just parades higher, and no citizen can stop that. But when it comes to the water rates, it's very odd how powerful the citizens become, as the constituent voters are surely the reason that water rates are suppressed—or so political voices tell us. The need to improve infrastructure to avoid jumping cancer rates and water illness should be obvious. We pay for that illness in lost productivity, hospitals bills, and mental anguish—all real costs not factored into our paychecks and wallets, essentially morphing into our own ULs. We must recognize that the arteries and veins of this nation are the pipes to our heart's core, which bring freshwater to the home and send wastewater to be treated. We must begin to ask our municipalities to raise rates.

If families couldn't have water and sanitation without higher water rates, wouldn't they pay without a protest? And if not, who was being represented when political forces continuously suppressed rates at HRSD? Were Virginia families so outspoken to groups like HRSD that they removed a politician from office and influenced others to keep rates below breakeven? No, the majority of the constituents don't vote. We can't blame their votes, although they are the scapegoats. In fact, presidential candidates harp on this vision of their own "New Deal" or public works adventure, but end up making no decisions. Might we know the reason why?

A suppressed water and wastewater rate will not allow the economy to recognize water value like the EPA has. As we continue to learn, misrepresenting water value by promoting a water gap will benefit the few and continue to weaken the wallets of everyone. Let's close the gap! I do *not* believe underpricing water is the collective interest of the US citizen casting their votes. I believe citizens are not yet aware of this long-term pain. This is our opportunity as Americans to rise up, take a stand as water value advocates, and lobby for water to be *Treated by America* rather than the French. If increases to water rates were made strategically over time through our government, we could adjust our families' budgets. We are talking about marginal monthly differences in our water bill: five, ten, or fifteen cents on each dollar of just those bills. We literally lose a couple of cents every time we break a dollar, as some change is never put back into the economy, identical to the way 10 to 30 percent non-revenue water is lost. Where is that dirty penny now, and what if we could capture all those lost pennies for our country? More importantly, if we dismiss this opportunity and let water become proportionately privatized, then we absolutely give up 10 to 13 percent to shareholders rather than to the pipes for our water bills. That 10 to 13 percent leak will go to some foreign corporation born in France. We will be paying for that French CEO's vacation in the Mediterranean. That CEO is on the same damn island you wanted to visit. Where do you want to be?

In sum, we are talking about our families' lifeline to prosperity. What do consistent rate increases really mean to our pockets? The price of everything else is rising, so why shouldn't our water increase with the cost of Skittles?

I encourage everyone willing to complain about an *accurate* water bill increasing from a local government's operation to first certify the very personal importance of their water and water quality. Just think about how cold, how hot, and how useless we feel without electrical power—as we learned that water makes electricity a valuable aqua-export. But after coming this far in waconomics, I believe we now know what water and the service of wastewater mean to our families.

During my time at HRSD, I confirmed that industry water rates do not exceed, match, or track aggressive estimates of inflation, nor compensate adequately for NRW. Simply put, there was a water gap in the water rates themselves. There are real levels of water inflation at work today, even involving the investment in the water and wastewater pipes.

We now know how rates become invisible concerns. But we don't know why and we don't know who is connected to these political forces.

Water rate increases are choked off by a political force until the debts from bond sales are too high, the loans too much to service, and the EPA consent decrees too overwhelming. Once the rates are held down long enough, the situation becomes dire, as it did in Flint, Michigan, and many other US cities. Like sellouts, the leasing of the municipality to a French corporation is offered as a solution—just for twenty or forty years. As soon as the contract is signed, water prices can spike and almost float. If enough of these leases happen, it will be all pain and no gain for the people.

It's no longer appropriate to follow where the money is counted. Instead, we follow who might be associated with the corporations who privatized and hiked prices.

CHAPTER 14

Wall Street's Pipes

RECENTLY, THE FIDUCIARY RESPONSIBILITY of the US government to create and maintain infrastructures like airports, municipalities, ports, and roads has no longer seemed like patriotic duty or priority. Fostering the country for greatness is in the past and the history here is changing. The allocation of funds is diverted by entitlement, subsidy, or welfare. This is more of the same cancer protocol, treating the symptoms and avoiding the cure. Every pattern points to increasing symptoms as history repeats itself there. The very pipes that make the United States of America great are subject to corporate takeover. In 2006, the Associated Press and *USA Today* noted that foreign companies were buying up bridges and roads. Water systems are next. It's already well-documented that private investors and speculators, like T. Boone Pickens and, allegedly, the Bush family, are privately buying land sitting on top of aquifers or privately buying up water rights. Oddly, treating and supplying water is a natural extension of that business we call ownership.

Would you rather pay a toll to your country or a foreign corporation? That's the question we need to ask. Would you rather invest in your family or line the pockets of CEOs with abnormally ballooning incomes? Fund the middle-class dream or the corporation? Who still offers a pension, and who started taking pensions away from their own employees after promising them a funded retirement: the US or their corporate partners? In the end, the longer the US avoids a true water value and pricing that value into water rates for agriculture, industry, and residents (AIR), the longer we create more symptoms, and a situation where *we will have to* invest in those CEOs and their salaries. Wouldn't you rather make the choice than be told later that you have no choice? Let's be certain of one thing: when we make this choice, we will get a lot more back in return, and no one can take those benefits away like they did yours or your parents' pensions.

During the creation process that started in 2006, the US has been parallel to the European Union. "Bankrupt" Greece or Spain are compared to bankrupt California, Illinois, New Jersey, or Pennsylvania. But the world still counts on the US. The only significant difference between the EU and the US is fiscal federalism, the "ability to transfer economic resources from members with healthy economies to those suffering economic setbacks," as Krugman and Obstfeld observed. In the US, they added, "states faring poorly relative to the rest of the nation automatically receive support from Washington in the form of welfare benefits and other federal transfer payments that ultimately come out of the taxes other states pay." Moreover, the EU is limited in its fiscal federalism, because at the time of writing, it has not sold its version of a US Treasury bill. The ability to sell a US bond is fiscal federalism. The debt from Treasury bills supports all US states. Fiscal federalism is also a cultural problem, because full sovereignty is required. The European Union will get there, but only after each country homogenizes into one EU culture and accepts that sovereignty. In the end, it will be about money.

In Detroit, default is a normal discussion and reality on the nightly news. Thanks to fiscal federalism, I can worry less about how that immediately impacts me in Virginia. However, the concerns are much more far-reaching in the long term.

How long can we subsidize our weaker states and cities, addressing just the symptoms of the water gap, before corporations find legitimate timing in taking over our most critical and patriotic responsibilities? The bridges and roads are sold out to immune corporate bodies. Toll increases are usually an indication of new leases for foreign ownership. Why couldn't we have done that ourselves? And unfortunately, the water systems of the US are already being privatized. These indications are called "price hikes." Rockland County, New York, knows this. But what we need to ask ourselves is, why can't we raise the water rates ourselves? That's a choice, and it's a choice we can make and benefit from.

When will this privatization practice become a pandemic to the masses? The Chileans had debt just like US citizens. The Chileans were just like Detroit, Illinois, and New Jersey, and water privatization was their answer. But Chile is no shining example for the middle class. Its history was an experiment by the leaders of privatization—simply placing families in a beaker and watching their reactions, with a thick wall of glass protecting the scientists.

"Water privatization has not yet reached alarming levels, but the most dangerous scenario is the tendency to push privatization to the extent that it replaces public water control, as witnessed in Europe," according to *Water, the Source of Life,* published in 2005 by Dr. Rogate Mshana. And as we are already aware, water inflation in Europe is closely associated with its dying population, as pointed out by its fertility rate—European countries can't afford as many kids. "Scarcity of water is used as a threat. It should be noted that even in the US, public utilities still serve 81 percent of the American population. In some states, provision bounced between public and private owners. Some of the companies were accused of supplying water only to the rich and some for neglecting maintenance and repair work. In the US, much of the current involvement in water is in management rather than outright ownership of water works." We call them operations and maintenance (O&M).

When states and cities are vulnerable to bankruptcy, so are the municipalities linked to them. Conversely, if the water or wastewater districts are likely to default or claim bankruptcy, so are the cities, and so are the states with large enough water districts. It is happening now! So what has a default created? Let's check out Birmingham, Alabama, as every minute that passes creates more distance from the actors involved with the crime. Major media groups like Bloomberg had very clear coverage of the truth in Jefferson County, and those pages have been pulled from the Internet.

Instead of focusing on thriving areas such as Hampton Roads, let's focus on the meat and potatoes of cities around the United States before the EPA days. Looking back provides a clearer view of what is ahead.

◊>$

Consider the old coal and steel towns of Jefferson County, Alabama, where life is simple and *industry* is now quiet and *underperforming*. Jefferson County is just like Flint, Michigan, where so many in Flint relied on the growth of Buick, Chevy, and GM carmakers. The industrial growth in Jefferson County increased the need for absolute water treatment and sanitation, *especially* after populations flocked to the jobs demanded by Alabama's coal and steel producers. Under corporate leadership, Jefferson County had its own obstacles relating to the debts of the sewer and wastewater sector, specifically those prescribed by EPA consent decrees. Jefferson County went bankrupt in the twenty-first century because of its *inability to raise rates over decades* and fund sewer operations. The rates were always contentious among the businesses and citizens, according to a report titled "The History of the Jefferson County Sanitary Sewer System." Perhaps resistance was formed in Jefferson County after a century of lacking connections, communication, and efficacy.

The actual water system was purchased from a private company by the Birmingham Water Works Board (BWWB) in 1951. The board became a public corporation, a quasi-status that serves many forms. The previous, private control of the water systems existed when so many industries like coal and steel were booming in Alabama. Bigtime steelmaker US Steel shut down its furnaces and steel operations in 2015, after a century of production—leaving the environment a sponge, and thousands without employment reaching for food stamps.

At the time of default in the twenty-first century, BWWB served approximately 25 percent of the state's population in a five-county area. Today, after the biggest financial crisis in municipal history, BWWB still operates as a public corporation—and its water bills are suffocating not only the residents, but businesses too. Decades upon decades under BWWB leadership created a bubble, which most certainly popped, and a catastrophe ensued. Instead of houses being overpriced, the pipes and their water rates were underpriced, blowing bubbles of debt.

Something interesting occurred between the EPA consent decree in 1996 and the years before bankruptcy hearings. Conveniently, after half a century of debt and compounded EPA demands, the water system was sold in 1998 to the city for $1 (or was it?) with the hopes of facilitating privatization of the utility. In 2000, it was sold back to the BWWB, and ratepayers were asked to pay $471 million plus years of interest, according to published reports. Pat Feemster, a contributor to AL.com, an Alabama online newspaper, typed these comments: "The public should not be misled . . . Birmingham does not own the BWWB. It is a public corporation built by board members, established under a charter granted by Birmingham and authorized by an act of the legislature. It is funded by 261 revenue bonds and ratepayers, not by Birmingham." Well, who was responsible for those bonds? Because those bonds and their compounding interest—a realized water inflation—was also responsible for residential water bills equal to $400 or more.

BWWB allowed this bankruptcy to happen on its watch. Debts created from the sewer system were clogged with EPA requirements and with sewer rates that did not track those costs. And all of this came from almost a century of aging and inadequate infrastructure. This was a private sector gig that failed miserably over time. In the end, CIPs for infrastructure and those debts, connected to unreasonably low water rates, compounded the problem.

Could the sewer system of Jefferson County establish and maintain the pipes to the tune the EPA found necessary both historically and today? No, Jefferson County couldn't keep up, and its water system brought down the town—literally. The county had and continues to have a decreasing job market, matched by falling wages as those industrial jobs no longer demanded a workforce. A county that started with water stamps now results in more food stamps.

The following evidence highlights in glaring fashion what political force looks like, as found in the Jefferson County sewer system historical report:

> In late August 1971, the County hired an investment banking firm to perform the ground work for the bond issue to finance the $30,000,000 program. However, that same month, President Nixon issued a wage-price freeze, and the Internal Revenue Service (IRS) informed Jefferson County in November 1971 that a sewer rate increase was forbidden under the freeze. The County was caught in an impossible situation: the EPA had ordered the County to clean up its waterways, but the IRS had forbidden the means by which Jefferson County could upgrade its treatment plants . . .

That occurred just as wages in Alabama began their decline, due to the effects of EPA regulations on industry. Buying power weakness and dollar weakness had just begun.

Now fast forward to the twenty-first century. According to Bloomberg News and writer William Selway, Jefferson County, home to Birmingham, Alabama, is the site of the biggest municipal bankruptcy in US history. EPA lawsuits compounded the county's fiscal woes. Violations of the Clean Water Act for untreated sewage discharge led to a consent decree to "repair and rebuild its sewer system." This brought about an interesting slew of bond sales and intriguing use of derivatives—so-called revenue bonds, backed by sewer fees instead of tax collections. What started out as a $555 million offering ballooned to a $3.1 billion debt for Jefferson County.

Sewer fees over ten years increased 400 percent. Raymond James & Associates was involved at first, but the SEC later identified JP Morgan as the "dominant provider." Here's a litany of affiliated complaints based on Wall Street's connection: allegations and convictions of bribery, cronyism, excessive fees, fraud, jail time by county officials, patronage, secret payments by JP Morgan, and SEC settlements with JP Morgan greater than $700 million. The reverberations continue today. "At least 21 people have been convicted or pleaded guilty to corruption-related charges in connection with the sewer construction and financing," Bloomberg News reported. "More than 95 percent of the sewer bonds were traditional fixed-rate securities. By the end of it, 93 percent carried interest that fluctuated along with the market movement. In 2008, Jefferson County defaulted on a $46 million principal payment on sewer bonds."

These are very similar practices to those of the Chilean "free-market miracle." When loans or bonds tied to a weak dollar get fixed and locked, the debts explode higher when the dollar moves higher, as happened in 2008 to 2009.

Was the Birmingham Water Works Board acting in the best interest of US citizens? Did the water or sewer system accommodate the long-term negative externalities of the coal and steel industries' footprint in the state of Alabama? And did the results cost the people billions of dollars in unrecognized leaks? We say ULs because no one is compensating the people of Jefferson County or adjusting their paychecks. There is simply no relief. It's also a crystal-clear example of water inflation upon our lives.

If water and wastewater services were a public duty operated at the city level, overseen by the federal government, adopted with the mindset of fiscal federalism, could the bankruptcy in Birmingham been avoided? All that was needed in Jefferson County was a water rate that matched the increased demand for infrastructure and a relatively equal payment by coal and steel producers when they were actively operating in Birmingham.

Why was a public water corporation allowed to max out debts without physical results? In the end, the need for sanitation and sewer systems was lined all the way to Wall Street. Neither the corporation nor the citizens gave the importance of water and wastewater an accurate value. Debt is a sour subject for anyone who bears the cost.

It doesn't matter whether BWWB was a private-sector gig for a century prior, giving coal and steel a hall pass on their tenure in Alabama; it doesn't matter the price tag they didn't pay the environment when they were there; it doesn't matter what mess they left behind; it doesn't matter that BWWB became a quasi-public corporation. No matter who we are and where we come from, we make mistakes. But the infrastructure was overlooked and the water rates were never addressed until it was too late. Officials put off price increases and gave the keys to the city to Wall Street to manage their water's money. Private banks got involved and went after profits and, in doing so, attacked the people, their families, and their wealth into the future.

These setups are happening elsewhere in our country. So it *does* matter if the banks take an extra 5 percent or 30 percent from our wallets. It's our water for AIR, and that includes the water for corporations and farmers who produce goods to feed their families. Since it's our water, let's keep our money on it. If governments around the world like China can print water by printing money, and the United States Department of Agriculture does it for Texas, then we can print money for infrastructure loans and address those loans with our money, paying water rates that make cents and sense to our families. You pay for what you get, until you lose control.

It *does* matter that a corporation, over decades of ownership, could create this catastrophe with Wall Street banks. It *does* matter that businesses and residents could not see the water value necessary to offset the long-term costs. It *will* matter when it affects you, but why wait? It's happening and we can understand it now. In general, water rates need to rise, but how can our *paychecks* rise to offset that requirement?

What does that mean long term for all US cities and states? What city is next and how much debt? Which municipality will be forced into new business practices and absolute privatization? The "too big to fail" banks working with the corporations will promise to bail out the city, but what if they do the opposite, like in Jefferson County? The standards of living have regressed there. The town is dysfunctional and food stamps are on the rise. There are similar stories in Chile or Bolivia's history. We have to wonder how many *water stamps* coal and steel were given decades ago for their short-term gains, and how many *food stamps* that industry creates today. We can still watch news reruns on the Internet of these stories. It's not about *what* we do with the money, it's about *how well* we implement the money. Which municipality will become the next Jefferson County?

As I edited these words, signs of future privatization were flashing upon the screens in Detroit.

Detroit and the Detroit Water and Sewerage Department (DWSD) are facing a public dilemma. Water rates exceed the income level of the lowest socioeconomic classes—they are now upside down like in Jefferson County or the Philippines. The water has been shut off, according to CBS. Will DWSD eventually be privatized and raise water rates regardless of income levels? What if we become Bolivia? Standards in Jefferson County and Detroit already are similar to those in Bolivia. After all, Birmingham and Detroit now own the scariest crime rates in the US.

That's exactly what's happening elsewhere. Right near the Big Apple, the results of privatization are being fought in Rockland County. However, the fight is *not* between Rockland citizens and their government, but between lawyers hired by citizens and SUEZ Water—previously United Water, a subsidiary of SUEZ Environment. Boiled down, SUEZ is a multinational corporation from France. Rockland County faces rate hikes from 28.9 percent up to potentially 50 percent. The most ironic issues with Rockland are just the costs of plans proposed for a desalination plant on the Hudson River. The costs of the plans themselves without any construction are equal to $50 million, according to published reports. The citizens of Rockland County will bear those costs. Why is desalination so interesting? That is another Chinese finger trap. More important is that the people of Rockland have to pony up the money to pay attorneys to fight SUEZ in New York. It's out of their hands now. When the State sells the country's physical assets—priceless ones like water—someone is getting sold like a slave to the lands they live on. It's about the color of money.

◊>$

In the meantime, anyone heard of the media covering failing infrastructure in New Jersey? Understanding New Jersey's crumbling infrastructure is no mystery. Take a look at their free-market solution below.

In Bayonne, New Jersey, privatization has surfaced in a tremendous way. Bayonne now has a long-term SUEZ Water contract. In July of 2012, United Water or whatever they change their name to, altogether a French subsidiary, announced a forty-year operation and maintenance (O&M) agreement of the Bayonne Municipal Utilities Authority (BMUA). Private investment made major gains with this agreement with unparalleled momentum. Because of the size of the utility and the length of contract, SUEZ's approach in Bayonne is acclaimed as a new business model. Although the operations are very similar to other contracts SUEZ has operated, this time the financing was accomplished through another party, Kohlberg Kravis Roberts & Co.

KKR will provide capital outlays for improvements to the outdated sewer and water systems and initially pay $150 million to the municipality, which removes a debt obligation for Bayonne Municipal Authority—a.k.a. bailing out the city. United Water's red-carpet cost for the citizens of Bayonne will be an immediate 8.5 percent rate hike, met by a four-year freeze, and then 4 percent rate hikes per annum for the remainder of the thirty-six years. In fact, liens were placed on peoples' homes for nonpayment when water rates increased four years later, according to NJTV News. Why would Bayonne not raise its own rates and abstain from relinquishing control to SUEZ? Was there a political force holding down water rates in New Jersey prior to the forty-year contract? Were all the voters—who now have no choice but to pay SUEZ a higher water rate—the constituent voice begging Bayonne not to raise rates? The success of privatization years later was aired by CBS in June of 2017, when major water main breaks disrupted Bayonne's water on a hot summer day, "leaving many residents without clean drinking water." The private water taker blamed "aging infrastructure and temperatures."

SUEZ has been a standout difference maker compared to the people of the City of Bayonne, right? Why do we need SUEZ again? Anyone can blame the weather.

Focus on the nexus that Jefferson County has with Bayonne—private banks that are "too big to fail." KKR is simply a team of ex-Bear Stearns teammates. And we know that Bear Stearns failed and needed to be saved from itself, later merging with JP Morgan, with the help of the US government.

The most standout expenses in Rockland County are the same new costs that privatization welcomes—shareholder profits. We have already identified the fundamental flaws regarding new and old pipes. We know water rates have not met EPA demands over time and that these water rates do not match the scope of the money lost from non-revenue water. Now citizens must pay the shareholders of SUEZ, who privatized their municipality. Any water district already in debt and privatized is now exposed to shareholders' financial obligations. Let's just say 10 to 13 percent is now part of a water district's bill.

That means Bayonne just added 10 to 13 percent to its citizens' tab and did not apply that money directly to New Jersey's infrastructure. It does not make sense, for the only reason the Jersey deal happened was that the municipality and state felt financial pressure and SUEZ offered a solution—the banks' support. They could have paid for the change themselves. It's about money, but doesn't the US print money for every other cause? In the end, Bayonne citizens will most likely pay much more than the cure this book offers. Some of those costs will directly line the pockets of shareholders and CEOs, sending them to the Mediterranean Sea . . . To that same damn island.

◇>$

What occurred in Bayonne will likely happen elsewhere, as EPA consent decrees are issued because of problems with aging pipes. This will wreck municipal bond ratings, making it increasingly difficult for local governments to issue debt to pay for upgrades. S&P ratings will decline, making financing harder to come by, interest rates higher, and more municipalities consider privatization—it'll be so South American. Sadly, this trend could have been avoided with steady rate increases applied over many years. Bayonne, Detroit, Jefferson County, or Rockland didn't adequately raise rates, and their citizens are now worse off than they would have been with gradual increases. Now some of them face significant water rate hikes for a system they no longer control—at least until the agreement ends.

So, why are rates held down? Are special interests intentionally feeding the influence of a lower water rate? As I discovered while at the Hampton Roads Sanitation District (HRSD), some of the reason is conveniently political: politicians don't want to anger voters with rate increases. But do the voters at the ballots have the facts? And could it be even more sinister than that? I wonder, could there have been powerful political and special interests at work, hindering Bayonne from rate increases sufficient to maintain the public's control of the pipes?

My experience at HRSD and my research into the international political economy cause me to *speculate* that there is a hidden campaign in Washington, DC, most likely served by special interest banks, energy, and water groups.

$$\lozenge > \$$$

The road map of Chile's "free market miracle" bears a striking resemblance to what is happening here in the United States. Do you recall how much growth occurred for the citizens in that country? Poverty increased 4 percent over a twenty-four-year drag. Privatizing water treatment and distribution facilities drain economic prosperity and undermine the American dream. Over time, water rates will increase exponentially, no matter who runs the pipes. So this becomes our choice: increase the price on our terms to reinvest in our country or lie down for a French corporation and give away our power. Once you get shareholders involved, safeguarding and running water utilities is no longer about protecting the public interest—it's about the reward to shareholders and to CEOs. The American dream is offshored and outsourced to a French party. Privatized water districts become profit centers, pricing water to meet the profit expectations of very few.

As the privatization model spreads, it will begin to look alarmingly like a national trend. Flint, Michigan, serves as a great example. There will be many more Flints across the nation—it's happening. There will be story after story of aging pipes that leak or have contaminants, which will result in more EPA lawsuits and mandates. Municipalities won't have the bond ratings or cash to make the necessary fixes. Political corruption or special interests can override the water and wastewater authority, keeping it from seeking higher water rates from businesses and residents, causing the municipality to run even deeper in the red. Drowning in debt, the water district will seek bond sales, yet support from the ratings agencies are bludgeoned by a dying city and a weak and unstable housing market. Bechtel, SUEZ, Vivendi and other big corporations will line up as white knights. Privatization will be their solution. Teams of corporations and banks will code arguments with economic salvation, claiming there is much-needed cash available in their briefcases and there are gains from the private sector's efficiency. Once in control of water supplies, corporations will seek to maximize their profit by hiking water rates. It's a dire but seemingly inevitable scenario, unless municipalities take a stand now and begin increasing rates to fix the pipes.

It makes no sense to *refuse* to raise water rates for your country's long-term health, yet bend over for a foreign company later after a water district is privatized. It's time to get realistic and savvy about our families' finances. It's time to put a moratorium on privatization and say yes to a public water value at government-run utilities. If we give the city a raise, then it will give both citizens and corporations a raise. As we'll soon see, it's a win-win-win.

If the arteries and veins of our family are owned by those who administer our blood, then we should ask ourselves, what does Wall Street want with our blood? What has Wall Street injected into our bodies that will alter our families forever and constantly make us pay? Blood is not thicker than water; it is water.

CHAPTER 15

The Candle of Profits Burns at Both Ends

FRACKING—THE EXTREME OIL and gas extraction method that involves blasting millions of gallons of water mixed with toxic chemicals underground at enormous pressures to break apart subterranean rock—has exploded in the last decade. More than 270,000 wells have been fracked in 25 states throughout the nation. More than 10 million Americans live within a mile of a fracking site. This means that 10 million Americans—and truly many more—have been placed directly in harm's way. Hundreds of peer-reviewed studies have connected fracking to serious human health effects, including cancer, asthma and birth defects.

Wenonah Hauter, *Ten Years Later, the "Halliburton Loophole" and America's Dirty Fracking Boom*

If everyone you knew peed in the pool, would it still be a pool? Let's find out and host an end-of-the-year pool party for a middle school in Oklahoma City.

Imagine US citizens as the students. Explain the EPA rules as the pool's rules. Visualize the pool as US lands and imagine fossil fuels, like natural gas, to be the refreshing pool water that we all love on a hot June day. Imagine every classmate who got a hall pass as oil and gas producers—and it was those students who were granted early dismissal and allowed first access to the pool. And *remember* that all the pee in the pool is produced water, with hundreds of hazardous chemicals in its constitution.

On August 8, 2005, the Energy Policy Act was signed into law by President George W. Bush and, by his side, former Halliburton chief executive Vice President Dick Cheney. Before this date, the EPA gave water a value in the 1970s—whether the American people got the memo and understood the EPA's impact is beside the point. In many ways, the Energy Policy Act pulled the rug out from under the water value established by the Clean Water Act and the Safe Drinking Water Act in the 1970s. It did so just for oil and gas production (O&G) and removed most of the recognized water value from their operations. The Energy Policy Act gave aqua-exporters like Exxon what we term the *Halliburton hall pass.* Cute as this might be, it's actually known as the "Halliburton Loophole." These classmates, the producers of O&G, were dismissed a little earlier than the rest of the middle school for the summer. They arrived at the pool before everyone else for the end-of-the-year school party, and that hall pass exempted them from pool rules.

The oil and gas classmates who used their hall passes were not leaving the pool. When this news hit the school, everyone in class wanted to get to the party immediately. At first, the hall pass drove O&G companies into the water, cramping every inch, making the pool less attractive to the rest of the school that was invited to the party *on time.* On the surface, it looked too crowded. But it was worse than that. While this was not spelled out to rest of the middle school, the O&G classmates dismissed early for the summer were told it was okay to pee in the pool so they wouldn't lose their spot when going to the bathroom.

If every inch of pool real estate was taken and used like a toilet at one point in the party by every O&G classmate, and the middle school found out, would the rest of the school still want to play in that pool?

The value of the pool to the students would fall. And that's exactly what happened to the price of natural gas following the Energy Policy Act in 2005. Too many O&G students joined the party, because a very sensitive barrier to entry, water value, fell. *So many* producers took the Halliburton hall pass that the supply of natural gas dramatically increased and the price fell—totally dismissing water value, the environment, and therefore *discounting natty gas well below its true market price.*

Like a mini-earthquake, water and hundreds of *top-secret* chemical combinations are blasted thousands of feet into the ground, infusing our water supply. And if you don't have to include the cleanup cost when leaving those chemicals behind in the groundwater, that saves time and money for oil and gas producers. Yet water forever marries with that chemical cocktail in the aquifers, and then it's used to grow our food.

◇>$

The initial ramp-up to natural gas prices in 2005 happened because water was no longer calculated in CPI benchmarks for inflation, especially food and energy. And we can etch the year 2000 in our minds, after the last CPI revisions in 1998 and Boskin Commission results in 1996. So the only way water could *really* express its value was through the price of aqua-exports. No one else was speaking for water's worth, and that meant that aqua-exports like natural gas had more value than ever before. After 1999, natty gas was speaking for both the commodity value itself and water's missing value. Because the CPI elected to communicate that our cost of living did not represent water and water-intensive food and energy, natural gas climbed in price.

Natty gas prices were inching higher and higher approaching 2005, but then the industry got a major hall pass—excusing their requirements to safely dispose of produced water. O&G producers who created toxic chemical cocktails at exponential rates were extended the subsidy of a lifetime. That hall pass removed a great deal of red tape and water responsibility from their balance sheet; therefore, operations became much cheaper and everyone from the O&G sector just jumped in the pool. O&G proceeded to erase the water value established by years of EPA momentum. Relinquishing sensitive environmental mandates from the EPA and ignoring previously suffocating costs to operations, the oil and gas sector was shedding its overwhelming responsibility to deal with drip gas, a.k.a. produced water.

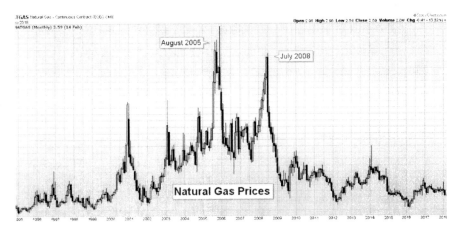

Chart courtesy of StockCharts.com

Natural gas is measured in MMBtu, which represents one million British thermal units. Within five months of the signing of the Energy Bill, the price of natural gas was peaking at $15.65. Afterward, nat gas plunged, and within fourteen months of this bill being signed, it was $4.40 per MMBtu. For years, O&G companies had needed to abide by certain EPA rules that made production very costly when you considered how they handled water pollution, a.k.a. where they peed in the environment. The Energy Policy Act removed a tremendous amount of responsibility for water cleanup and therefore operations became way too easy—no one at the pool party had to get up to go to the bathroom. The only time comparable prices were ever resurrected since occurred in 2008, following the droughts which began in 2007. But it was a *lower* high. Fortunately, O&G students knew that they could *free ride* on one of the biggest upstream cost to their productive profits—water—by leaving underground water sources with drip gas, basically peeing into the pools of US backyards. Natural gas could afford to be cheaper in 2008, even when the world was on fire and short water. Why? Because its most expensive cost, handling produced water, the overwhelming liability of all operations, was drastically reduced as the EPA amendments were modified and gave O&G an exemption from the rule.

As I edited this book in 2016, the price of natural gas was devastated relative to those spiked highs in 2005—*90 percent cheaper.* Think about the number ninety. It represents a huge discount to oil and gas production and a major accounting trick for them, wherein the social costs of produced water leaks onto the families and their home values. Ninety represents the percentage of all water required for the food and energy component of agriculture, industry, and residents (AIR). *Headline* inflation represents this food and energy and, more importantly, 90 percent of our global water consumption. Ninety represents the volatility the Federal Reserve argues is the problem in the CPI and PCE, the reason it instead uses the *core* inflation. Based on *core* inflation, 90 percent represents what the Federal Reserve removes from calculations for our COLAs, our social security payments, and how we compensate our firemen, military, and teachers for their time. And much, much more. So, what does the Halliburton hall pass look like regarding the Safe Drinking Water Act?

> The federal Energy Policy Act of 2005 amended the Underground Injection Control ("UIC") provisions of the Safe Drinking Water Act to exclude hydraulic fracturing from the definition of "underground injection." The objective of the Federal UIC program is

to protect underground sources of drinking water from contamination by underground injection of hazardous and non-hazardous fluids. However, protection of groundwater resources during oil and gas extraction activities is a responsibility of state government.

NYS Department of Environmental Conservation

But the Halliburton Loophole wasn't the only fracking enabler in the Energy Policy Act. The act granted the Federal Energy Regulatory Commission (FERC) sweeping new authority to supersede state and local decision-making with regard to the citing of fracked gas pipelines and infrastructure. It also shifted to FERC industry oversight and compliance responsibility for the National Environmental Policy Act of 1969, another key law. This was akin to putting the fox in charge of the hen house.

As it stands, FERC is entirely unaccountable to public will. It is unaccountable to Congress and even the White House. Commissioners are appointed to five-year terms and can do as they please. Until a law reigning in FERC is passed, the commission will continue to act as a rubber-stamp for the fossil fuel industry.

Wenonah Hauter, *Ten Years Later, the "Halliburton Loophole" and America's Dirty Fracking Boom*

And what about the Halliburton Loophole regarding the Clean Water Act?

The federal Energy Policy Act of 2005 defined "oil and gas exploration, production, processing, or treatment operations or transmission facilities" to include all field activities and operations related to these facilities "whether or not such field activities may be considered to be construction activities." The effect was to exempt well site activities that disturb one or more acres from the Clean Water Act's requirement for National Pollutant Discharge Elimination System.

NYS Department of Environmental Conservation

Most economists would opine that the price of natural gas plunged since the twenty-first century began because of the innovation of technology. On the contrary, *the rally in US domestic energy production* could not have occurred without the reversal of the protections safeguarded in the Clean Water Act and Safe Drinking Water Act. The abundant production of natural gas and the "energy boom" *cannot* take place without very liberal laws regarding the water supply and produced water. Otherwise, the costs would be so much higher, only allowing the most efficient producers into the market, driving up the price of O&G. But we know that is why prices plunged, because the number one barrier to entry was minimized relative to costs. Water inflation again was hidden from our eyes, but we pay for it in our bodies.

Fracking's proprietary chemical blend is legally protected from the public. This is *sketchy*. It's like a cocktail prepared for you by a stranger, with date rape drugs in the mix that you don't know about. No one can dispute what's in fracking's cocktail, and the few of us who know cannot protect anyone else.

Heading back to the Golden State, we recall that California is the world's fifth largest supplier of food—if it were its own country—and the third largest producer of oil and gas for the US. Remember that California has produced 244 million barrels of oil in a year and it needs 10.5 barrels of water to make each oil barrel. So we can simply multiply the barrels of oil times 10.5 barrels of water to figure out just how much hazardous water is used to grow our food. Water remembers, and so does our body when we consume that food.

Now let's go to the grocery store. Imagine the lights, the cement walls, the shopping cart itself, all the fruits and vegetables, the meat and dairy, the packaging made from oil and gas, the diapers, the pharmaceuticals, the chemicals for cleaning, and the bank accounts supplied with our time—it's all water, isn't? What if *we* could get a Halliburton hall pass, and a majority of the dead water from production, contaminated water, the slaughtered blood water, the fertilizer runoff, the human-made chemicals, the discharge from drugs, basically the negative externalities and therefore the unrecognized leaks, was discounted and removed from our bill? *Would we save substantial money* at the grocery store, in the short term? Yes. Because the prices to live and thrive were abnormally too low. But we would realize the social cost to our family is too high when unrecognized leaks (ULs) go into overdrive. What's the reward for the shortcut fire sale? Would we be an improved version of ourselves in the end? Would we be innovating? Nope, we would be holding back change and punishing our wealth and health later. That's the only constant. Something will eventually give, and what we mean by "give" is "take," from us. The future is unknown for that hall pass.

⬗>$

The downside to the Energy Policy Act is that when we remove decades of regulation, natural gas becomes falsely underpriced. The public absorbs the social costs and ULs. The industry was allowed to achieve more production with very little consequence to its business model—a negative externality (NE) that becomes a major loss to the people. In this light, oil and gas companies received incomparable subsidies. All the while, the world interpreted the new supply of oil and gas exports in the twenty-first century as our rise back to prominence. The US finally overcame the decline in exports that had plagued it since the 1970s, but at what cost?

When something is too cheap, it becomes a steal, and that causes everyone from the power companies to the consumers to buy more than they should. Burning oil and gas is burning water. The Halliburton Loophole to the EPA acts promotes an investment model to burn water through fossil fuel production and simultaneously invest in physical water elsewhere (as T. Boone Pickens has). The candle of profits burns at both ends. Remember, while the financial returns versus actual production are said to be on a "treadmill" for oil and gas, *the water is certainly not*. The sacrifice created by the excessive water demands and pollution in O&G are for short-term gains in production. So if we're exhausted by being on the *treadmill*, then realistically, once we stop running, we will go backwards for just two fossil fuels. As a result, all classes will regress. The underlying issues here are two.

One: water has no voice to be heard and therefore the people are bludgeoned by the impact of produced water on the environment, due to an irrational long-term subsidy that leaks into the environment and our families. Two: when water is taken out of the equation, we have no control over the price of natural gas, which greatly increases volatility.

Volatility is the one thing the Federal Reserve is trying to remove from our economy. *The way the Federal Reserve has solved this problem to date is by removing more and more water from inflation numbers until the noise disappears—a.k.a. the volatility.* But that does not remove the higher prices of healthcare we will pay, because no one is speaking for water quality in our paychecks, or protecting our families from its lack. In the end, if water quality goes down and the US government's inflation stays the same, we are getting robbed.

The price of fossil fuels swings wildly when water is not included in the equation for O&G producers. For example, in 2008, the prices of natural gas overshot to the upside due to worldwide drought—oil prices did so as well. Meanwhile, without adequate valuation of water rates, the price of oil and gas can also fall far below market value—creating a long-term systemic risk to the prices of fossil fuels and our water supply. This creates a *market failure for energy* first. When oil and gas prices fall so low (like they did in 2016 and 2017), it's giving water a hall pass, dismissing it from school and not using its power—or rather, abusing its force. But it does not excuse the thirsty classmates, us, who need quality water to thrive. Not only does the O&G business model not recognize water value, but it creates a *market failure for water quality*. Additionally, at such low prices, producers of oil and gas can't keep producing, so they shut down and lay off people—as was the case in 2016. The price of oil and gas being so low pushes competition out of the market, forcing mergers and acquisitions, and strengthening the O&G sector by the day, even as it narrows to a controlling oligopoly of very few producers.

The oil and gas industry might want us to think that abnormally cheap natural gas prices are ideal conditions for the industry long term, but they are not. *They are scary.* There is no such thing as a free lunch, and we are going to pay. Cheap prices make costlier renewable energy platforms like electric cars, solar panels, and windmills less attractive— which prolongs their mass production. Cheap prices allow for the investment in switching from coal to natural gas, bringing new users online and locking up manufacturing plants to specifications that run those plants on natural gas. But once everyone and everything is locked up to the gas lines, the price will rise for the people—especially *if* water is privatized to the point that ownership makes it a commodity.

What happens if we ignore all these circumstances? Once water is less of a public good and more of a private one, game over for the citizens. If natural gas is everywhere, and we remain dependent on fossil fuels with abnormally cheap O&G prices at the beginning of the twenty-first century, then once water is predominantly privatized and then outwardly commodified like gas station lights, it will drive these oil and gas prices to nosebleed levels. And then every basic good will follow that price higher. But the coaches at the Federal Reserve will point their fingers at that volatility, and unfortunately try to remove the impact of more water from *inflation readings*, and therefore not allow our COLAs and wages to increase or adjust to price hikes. The water gap will widen in our wallets, and water inflation will rebel. This is why we must act! Even if short-term noise seems to disprove wa-conomic theory, over years and decades without any significant changes, this will be proven true. Unless, of course, we recognize water value now. If we ask our government to take control of water rates to agriculture, industry, and residents, we will get to enjoy the *AIR*. If we don't, then we'll get choked off by it.

Prices of oil and gas surge and sink. One reason the price of energy can swing wildly is because there is no true water value baked into regional production costs. If we gave water a value regionally for the energy producers through higher water rates, we could put a price floor in the actual production of energy itself. The prices of O&G have no business being as cheap as they were in 2016. The cost of crude has no business being $27 per barrel; the cost of natural gas has no business being $1.61 per MMBtu. Why? Because we must consider that water quality is being destroyed in the production process. *And* because the dollar amount that is lost in our economy the longer oil and gas prices remain that low is a major opportunity cost for energy producers— values bigger than many countries' total annual GDP, and all those profits lost by corporations.

The beacon of light for the economy, the people, and the corporations *is* to recognize water value. Once producers transparently digest the water value through water rates, the price of oil and gas is protected in so many ways by a floor of new operational costs. Hence, the price of oil and gas must rise. That's a great thing for O&G producers—they get to increase prices. If we increase the value of water through the government's water rates, the price of O&G *cannot* violently swing up and down. The price of oil and gas would not be allowed to fall below market value and not be allowed to scream higher the way it did in 2008. Because the value of water would be acknowledged, creating transparency, the O&G producers would be assuming costs, meaning little to no fluctuations and surprises from water later.

Transparency about water value and pricing water rates accurately through local government makes the impact of floods and droughts less relevant—since it is already priced in. When we recognize water value, we *actually* remove that volatility. And when we increase water value to the industry, the industry can charge higher prices—making up the loss from higher water costs. We can stabilize oil and gas prices through water value and remove the volatility for the coaches at the Federal Reserve. So let's forget about regulations for the sake of this conversation. If the price of oil is rising due to persistent droughts, a higher value for water can prevent it from overshooting.

This price ceiling comes from knowing what the full cost of the production process is, as priced according to its US state (or federal region on government land). The problem with overshooting energy prices is that it destroys demand. During demand destruction, actual producers thrive in the short term, then they become devastated as prices collapse. They chase falling prices, losing money on the way down. When we put water value into the oil and gas sector, we install a floor in prices, not allowing oil and gas to collapse any lower. Collapsing and overshooting energy prices are what the Federal Reserve wants to avoid. Instead of omitting water to get rid of that volatility in our inflation numbers (the way it has been done by the Fed), we need to bring water back into the equation. The key to economic growth, stable socioeconomic classes throughout the world, and opportunities to innovate is steadily rising inflation—as highlighted by corporations who make things. By placing water value in the metrics for all oil and gas producers, we are reducing the volatility of the globe's most addictive resource. This creates consistency—and consistency is king. This is something the Federal Reserve coaches want, so let's give it to them. Come on, Fed coaches, embrace this!

When we create water value, we allow consistent growth, which will stabilize social class no matter the city and offer the inherent opportunity to innovate to the entire human race (more on this in the pages to come). This creates transparent and beneficial inflation, long-term, for corporate growth. We are not socializing any resource; we are simply *accounting* for water and thus stabilizing the growth projection of economic outlooks, *rather* than overreacting to its hiccups. Producers of oil and gas *can* keep producing this way, *rather* than shutting down and laying off people during demand destruction. Moreover, the industry will *not* be controlled by only large producers—an oligopoly—who can dictate prices, like the Seven Sisters of Standard Oil. Recognizing water value would indirectly cause the oil and gas producers to innovate technologies which would prohibit pollution; when something is more valuable, it will *not* be treated the way the Energy Policy Act granted.

All and all, water value equates to less pollution, forecasted and managed prices for the energy sector and other industries, increased prices for the producers, consistency for business models, less volatility for global pricing forecasts, and stable pricing for a more all-inclusive approach to measuring inflation for the Fed. It's a win-win-win, because government officials, corporate partners, and we the people all benefit.

◊>$

A common misconception about politics regards taxes. And we can count environmental regulation as a tax contribution we all pay. Republicans want to reduce the size and breadth of government. They believe fewer taxes (capital preserved) and more pro-business leniency promote substantial economic growth. Democrats want more government programming. They believe more taxes and more of a say in how government operates and oversees the private sector bolster economic growth.

But how we look at taxes is the question. A tax is a negative—an upfront cost. It is a confiscation of our assets or income for a general fund. That fund could go to many channels within federal government agencies and be applied within our very own cities. At first, taxes obviously feel like a negative externality, where our paychecks have been wronged. Furthermore, they feel like a social cost we only mitigate through the election process. However, when we compound our taxes together, they are multiplied, and a negative times a negative is a positive. Strings of negatives can produce a positive effect not always apparent. Initially, taxation seems like a duty that unfairly weakens us individually, but is it?

For example, in the US, endless tax dollars are utilized for our biggest expense, defense spending. Those tax dollars are invested into military bases within the US and abroad, only to become positives. How do all those withdrawals from our paychecks become positive sloping checkmarks for our family long term? Despite our individual socioeconomic standings or our tax bracket contributions, all citizens were taxed and all citizens financed those military bases. And all US citizens benefit from the security a US military branch offers each city.

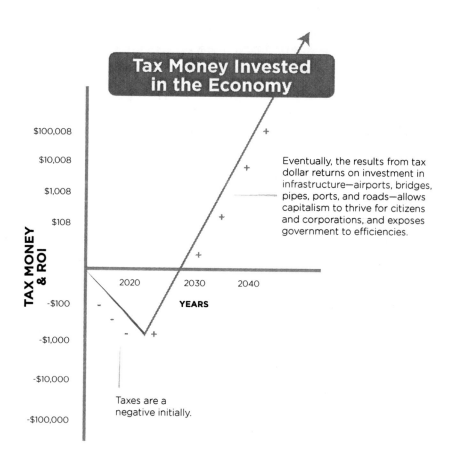

Tax Money Invested in the Economy

Eventually, the results from tax dollar returns on investment in infrastructure—airports, bridges, pipes, ports, and roads—allows capitalism to thrive for citizens and corporations, and exposes government to efficiencies.

Taxes are a negative initially.

In the end, the tax dollars are an investment in our military, which protects us. That is the spillover benefit or return on our investment, a positive externality (PE). Tax dollars become positive sloping checkmarks for families long term. Tax dollars compounded become an investment in our bridges and roads, which helps us obtain goods and services and get to places faster and safely, another PE. Tax dollars are an investment in our agencies, which look after US citizens after too much immigration, again a PE for those who believe in certain immigration policies. At first, it's a negative externality (NE) on our wallets and we bear the social tax, as seen in the chart above where the checkmark begins and then dips. Ultimately, however, once tax contributions are multiplied, negatives from our accounts become positives in the money quadrant above, and those taxes lead to an exponential return on investment. Look at taxes like checkmarks. In the short term, they are proven negatives, but the upside to taxation is investment in our economy into perpetuity, *if managed correctly.*

Economists and political parties believe the military is a public good or public duty for our country. Water has been one reason for wars, and water is a majority contributor to all the materials in the manufacturing process of every good for defense. So while it is suggested that Democrats' programming too often promotes "public goods" such as infrastructure, Republicans absolutely need water to fortify public goods like defense. So under these conditions, water and the pipes are not a public good, they are a *public duty for all political parties*—a national good for the manufacturing of defense. It's time we look to the artwork of that checkmark to dictate the times. Let's make some checkmarks for Democrats and Republicans! We can use $1 in water investment today (an increase in water rates) to get back $8 in financial returns for our businesses and families.

$$\Diamond > \$$$

What if I told you between 2004 and 2014 specific water sectors outperformed the general stock market as well as benchmarks for gold and energy ETFs? Would you believe me? If water is doing it for the private sector, why can't water do it for the whole country?

Our focus in waconomics is price—which means numbers. "Figures don't lie, but politicians do figure" is an appropriate adage here. I have no political party and have greatly admired both Democrats and Republicans. Even more important than their politics or religion are their numbers, as the office of presidency sheds truth when we evaluate the number related to time and money. If Republicans believe that less government and less taxation benefit the people—especially when running the free world—then let's objectively measure their belief system by numbers rather than subjective speeches. Waconomics consistently aspires toward goals that appeal to both sides of the political aisle, as well as solutions that offer a compromise unlike anything else in Washington. By pursuing such, I feel as though I have earned the right to offer a page or two of subjective political groupthink without losing the respect of readers. So let's step outside the box for a little political fun.

If the Republican argument is that the economy strengthens under Republican tenets and citizens of the United States are better off, then let's dissect recessions and whose watch they spend more time on. Since the first version of the Clean Water Act was born in 1948, we will use those years moving forward.

After the inauguration of President Truman, a Democrat, Republicans dominate in terms of recessions. They are clearly leading the time measured in months where the economy is contracting in economic output and prices are falling—and if you are pro-business, contracting prices are counterintuitive, *unless there are profits only few see.*

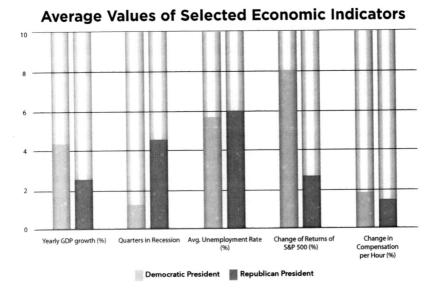

Average Values of Selected Economic Indicators

Source: Binder and Watson, 2014

Republicans lead history regarding recessions. Staying true to form, let's supplement this conversation by acknowledging that a Republican congress helps Democratic presidents balance economic growth during their tenure. And both political parties have argued that the previous administration's policies have delayed reactions. But the truth is a lot more difficult to swallow. Why is that? Because executive orders have expedited expensive wars on foreign soil without causing a hiccup in time but blowing a balloon in our national debt. And defense is our biggest expense.

But a simple look at the nature of taxes is noteworthy. If you take a checkmark away, you don't check off a box, you don't get it done, and something remains outstanding without completion—consistent growth. Taxes are checkmarks that become positive.

When we withdraw investment from a business, that business will most likely regress and the profit margins shrink. The same is true for the country. While Republicans argue that less government and less taxation is a benefit for the economy, then they are removing their federal, state, and local checkmarks. Republicans feel that when you allow corporations to have more access to their own profits, the US economy benefits. Well, if so, then recessions should be led by Democrats, who put more money into government and more money into US agencies that require more taxes—a.k.a. negative externalities, which are spun into positive externalities. *But Republicans prove the opposite, once we understand the process of a recession.*

If Republicans allocate more resources to private firms, by allowing them to keep more of their profit or by giving tax breaks or subsidies, then the argument that the economy is better off is convoluted. Historically speaking, when Republican presidents approach fiscal policy as they have, they remove checkmarks (taxes), or investment in the economy (government expenditures). And the number of times this has happened on a Republican's watch proves that less taxation and more pro-business dollars—a positive number for businesses and a negative number for government—when compounded and multiplied, equal more recessions—an upside-down checkmark.

Withdrawing investment in the form of taxation and offering private businesses more opportunities to grow (subsidy and tax breaks) actually creates more contraction for the people, as seen through recessions. When corporations have to pay fewer taxes, they invest less into their country for government spending. Government expenditures are one of the largest components of GDP, so when you reduce the funds (tax money) for government spending, you automatically shrink.

As a nation's GDP shrinks—especially for two consecutive quarters—it instigates recessions. Moreover, prices of all goods and services tend to fall in value during recessions, and corporations cannot give their employees raises when prices are falling. Falling prices are the opposite of corporate desires—remember, corporations need steadily rising prices to give their employees raises. Corporations actually lay off people under these conditions. In sum, people don't get a raise under recessions. Moreover, they sometimes take a pay cut, sometimes lose a pension, and often become unemployed. So the tax money the corporation doesn't have to pay under Republican leadership or the business subsidy the corporation receives from Republican programming *almost never makes it into the hands of the people, a.k.a. the economy. It only benefits the CEOs and the shareholders.*

Because prices are falling in a recession, the corporations don't even get to do the most important part of their business model—increase profits through steadily rising prices, a.k.a. inflation. In a recession, we have deflation, not inflation. So a reduction in taxes for the sake of the US economy doesn't pay the country very well if only a few people benefit. The masses do not get a raise, and we stagnate until we reinvest in the country. Meanwhile, the people of the country are incrementally held down in waiting.

$$\Diamond > \$$$

We can give all employees a raise in real wages by closing the water gap. When we do this, we can build the middle class and, in doing so, elevate socioeconomic classes into new income brackets, without taking away from more advantaged groups. It is scarcity thinking to say that we must take away from one group to benefit another. We can all rise with water value.

While innovation is the key to the future, we can better invest our dollars in infrastructure through recognizing water value first. We should pay for only that which we use. And as we'll learn, we can close the water gap and remove ULs by offering a higher water rate through local government, keeping water municipalities run by government, making water districts national treasures, and including a water quality index in our inflation benchmarks.

We don't have to increase or decrease taxes, we don't have to entitle or subsidize, and we do not have to create more free lunches. We can do the opposite, making the economy grow through water, the highest upstream input to productivity. If we want to change the future, we have to start at the top of the production pyramid to get to the base of population.

If we are here to innovate and improve humanity, then let's find the cure. Treating the water like *our superhero* and letting it become something greater *allows us all to benefit. It makes us more powerful* rather than incrementally held down everywhere.

Table 1.2: TEN GLOBAL RISKS OF HIGHEST CONCERN IN 2014

No.	Global Risk
1	Fiscal Crises In Key Economies
2	Structurally High Unemployment/Underemployment
3	Water Crises
4	Severe Income Disparity
5	Failure Of Climate Change Mitigation And Adaptaion
6	Greater Incidence Of Extreme Weather Events (E.g. Floods, Storms, Fires)
7	Global Governance Failure
8	Food Crises
9	Failure Of A Major Financial Mechanism/Institution
10	Profound Political And Social Instability

Source: Global Risks Perception Survey 2013-2014

In 2014, the World Economic Forum, the University of Oxford, the University of Pennsylvania, Swiss Re, and many more got together in Switzerland to discuss "Global Risks." While water was number three on their list, every other "global risk," as we now understand, is actually related to water.

So what are we risking by not lobbying for water value now? *Everything.*

CHAPTER 16

Win-Win-Win

ACCORDING TO GENERAL ECONOMIC theory, a market failure in airports, bridges, ports, and *now* real estate justifies government intervention. We call this action our *public duty*. So that's what we are going to do as citizens—intervene. We are going to intervene through price. We are going to do this for our families, for the US, and for global prosperity. As Krugman and Obstfeld put it, "The theory of the second best states that a hands-off policy is desirable in any one market only if all other markets are working properly. If they are not, a government intervention that appears to distort incentives in one market may actually increase welfare by offsetting the consequences of market failures elsewhere."

There is so much evidence of water's force and its connection to our economic regression. If we ignore it, our families will suffer slowly. The control of our water and wastewater will determine the fate of our dreams. There are water markets within the United States that are physically crumbling and financially failing. Privatization wants in. Private-sector water rate hikes shock us, and the downstream implications would become financial terror. Permanent price hikes can affect everything about our economy. If water is incrementally privatized and eventually commodified, like gas station lights, then the cost of living will be dramatically higher—*way too high*. If the city ignores its public duty, the middle class won't just regress as it has since the 1970s, it will evaporate.

But if we're aware that prices for water rates are rising through the city, there's no surprise and no terror because we have time to prepare. We have time to reposition our monthly budgets, and we can create a game plan for the initial rise in the cost of living. There are huge benefits behind this change. With awareness, *we get to juggle time* and control our destiny, on our terms, for our families. We should not be inspired by fear; rather, we should endorse this opportunity with foresight now. I am selling hope and I want you buying in today.

Since water quality is directly correlated with our food quality and healthcare costs, I believe that improving the quality of water in the economy will strengthen our families' financial wherewithal. Because of the insurmountable gap in water leaks and loss (non-revenue water), I believe improving water efficiency can rebalance our wealth and fortify real purchasing power for every North American—and eventually across the free world. Using water as *our greatest tool* can benefit existing areas of weakness and pockets of debt, while also improving total efficiencies. Water can subsidize weaker industries that have a social benefit despite their deficit behavior. Let's use Norfolk water and the libraries and police departments as a brief example.

Norfolk's water is rated one of the best tap waters in the United States. During my quick tenure at HRSD it was in the top five, and sold to other parts of the Hampton Roads area. Norfolk actually allocates its water revenues to the libraries and police force, who cannot afford to run on existing monies. This is a form of *local fiscal federalism* through water rather than US Treasury bond sales. Remember, fiscal federalism is the "ability to transfer economic resources from members with healthy economies to those suffering economic setbacks," through bond sales. Most importantly, this is an example of how water revenues can subsidize non-profitable but socially beneficial institutions. Yes, water is lifting up our libraries and strengthening our police force. Water is much more than a public good: it is our superhero, and we are water's sidekick, performing a public duty—our civic duty.

Why is Norfolk's water still underpriced if it can afford to fund other deficit operations? Because Norfolk also serves as a climate change laboratory. Remember, it has a flooding problem that stems from a sewer problem. Increasing the water rate to keep Norfolk, a city similar to New Orleans, above water is an expensive task. This financial contribution is going to require higher rates to prevent water inflation from depressing the value of Norfolk's real estate.

◊>$

All the while, allowing the water privatization model to incubate before water's full-blown commodification gives birth to more hidden water inflation. Under commodification, water value would eventually be publicly acknowledged by force—with water prices having the same glowing lights as gas stations. With more incremental ownership of water and new private leases for operation and maintenance of water districts, the risks of commodification grow, where water becomes a privilege and *not a public good*. In the end, this would be financial terror inside our wallets.

The water gap is partially the result of corporate participation, inviting private businesses to seek out profits by leaking social costs upon us, a.k.a. our unrecognized leaks. With enough city privatization and water rights purchased, the fact that we have not yet recognized water value could spike the prices of energy and food, even worse than it did in 2008. As history indicates, there will be just as much non-revenue water (NRW) with privatized water and wastewater operations. There will also be more leaks to shareholders in a privately-run water municipality rather than a financial return to the city. There are, and will be, plenty more price hikes on water bills from here until the end of history, as long as water can be controlled privately. This will offer no downstream payoff to the overall economy, because salary and wages would lag too far behind understated government inflation, and because corporate investment does not maximize returns to the city. The payoff to shareholders rather than the city and the *history* of privatization's lack of improvements to infrastructure leads to a loss, while customers or ratepayers only see hikes in their water bill. In the long run, privatization becomes a net negative investment both structurally and socially.

If the public is aware that water value is *now* a recognizable cost for the economy, water inflation can decrease dramatically. The water gap would close, increasing real purchasing power for all classes. Let's go through the process of publicly recognizing a water value and having that value increase transparently. This is the evolution of a new domestic fiscal policy toward water value. First are fiscal realizations, then a line-by-line economic tour by expectation. Theory is not absolute, but the same old economic patterns are ready for a twist. In the end, there is an international component which takes a tempered view of fiscal federalism and extends the theory of the second best to springboard global growth. But domestic fiscal policy comes first.

If the water rate is increased publicly rather than hiked privately, the cost of running all industries and residencies will rise, but new, efficient pipes will *join* the equation and become a game changer (less NRW). If prices can rise (the water rate), operational costs will increase for agriculture, industry, and residents (AIR), increasing the prices of goods and services charged to the producer and consumer. The prices of goods will reach their true cost structure which, believe it or not, will benefit our families long term. This will bring more revenues for corporations, and increase tax receipts from the corporations and citizens without ever increasing tax rates, which will generate budget surplus potential. Commodity pricing will increase and real inflation will emerge, raising interest rates. Ultimately, nominal wages will follow revenues upward and savers will be rewarded with higher interest rates in the bank. Not only will interest rates' rise be a positive externality for savers, it will curb inflation and campaign to investors a faith and promise in the municipal bond market—a win for the city and state. If the bond market is strong, citizens can make money investing in utilities because the yield is promising. Moreover, the corporations will have higher revenues and will not depend on zero percent financing and other life-support from central banks. Solutions are a derivation from the pipes.

⬦>$

So, how do we get cities, states, and the country on the right track? Create a *National Public Water Mandate*, a.k.a. water value policy, whereby each state must create target level water efficiency rates to meet existing CIPs (capital improvement plans) and future EPA demands. While we realize the EPA can be politicized, we remain prudent in keeping with our goals for water value:

⬦ **Water rate increase—city/county led:**

Increase the water rate federally through city- or county-managed and operated water districts. Create a National Public Water Mandate whereby each state's cities and counties are federally required to create target water usage rates for their CIP and EPA needs. The water rate increases will be based on metrics such as "headline" inflation and variable water inflation adjustments.

⬦ **National treasury advocacy:**

Water is a national treasure. Establish city-run water and wastewater districts as national treasures. Historic buildings are no different than our cities' pipes underground. The pipes in our buildings made US history rich with possibilities. Maintaining and upgrading all pipes is a credit to our heritage and future. Let's not hang onto the facade of a great nation through our history; let's be a great nation through action.

◆ Moratorium:

Place a moratorium on new private leases for operations and maintenance of local municipalities and allow existing leases to expire.

◆ National water quality indexation:

Place a national water quality index inside the CPI and PCE, based on random readings throughout each state and near industries.

Taking such broad actions will fortify a shift in this country, a projection that places the United States on the highest mountain with the brightest light beaming over oceans. Leading the world was once a part of the American dream. Being the example filled our hearts and our wallets. This will not just happen by itself. Forces to maintain the status quo run broad and deep. What is needed is a campaign of awareness to spread the word about water value, water inflation, unrecognized leaks, and our water gap. Change begins with awareness. As we believe that recognizing water value benefits us, conversations and then acceptance become key ingredients of our families' long-term wealth and health. Without a populace-wide belief that giving our water value a raise is a must, we will never overcome the biggest objection in the room: *Voters won't keep political figures in office once they increase water rates.* If we remove that objection, then we can give ourselves, our country, and our corporations a raise. Allowing our city-run municipality to be a little more aggressive with water rates and wastewater rates is key to our wealth and health.

What's in it for everyone? Let's find out through the possibilities:

1. **Ask for just a little more government involvement upfront, then let us profit, and in the end, give less back to government.** Invite your city-operated water and wastewater district to increase the water and wastewater rates to AIR. If water value increases by perception and above previous rate increases, this will fast-track capital improvement plans and the EPA's mandates necessary for the public works. Initially, we pay more to the city *(more government)*, experience the benefits herein, and from the benefits reduce entitlement, intervention, and subsidies *(less government)*, which increases the likelihood of a federal budget surplus.

2. **Give the corporations what they want, but on everyone's terms.** Corporations seek inflation and steadily rising prices, while deflation is counterproductive to earnings and the perception of economic health. Let's nurture corporations' inherent desires. The US should bring awareness and campaign to the public for future price increases for water rates across all users. Increasing water value with higher rates causes the operational costs to rise for AIR. And when costs increase, prices of goods and services can slowly and justifiably rise and corporations can grow organically with rising prices and *real* rising inflation. Corporations can increase prices and reduce their reliance on discounts, tax breaks, and water stamps.

3. **Why use something 150 years old?** Once realistic new water rates properly fund the CIPs (and EPA mandates), municipalities will generate more water

revenues and then reinvest them to expedite the replacement of expired, broken, and leaky pipes (reducing NRW immediately). Water districts can simultaneously introduce proactive water metering systems and digital monitoring technologies for breaks. This prevents the water gap in the long run and contains our unrecognized leaks (ULs). Physical water inflation from lost water dissipates, yielding long-term efficiency, immediate social benefit, and positive externalities (PEs). What happens then?

4. **Gain 10 to 30 percent more money per day.** Historically, the US has been losing roughly 10 to 30 percent of its most upstream input to productivity each day—a.k.a. NRW. Now imagine that percentage to be money gained, and visualize the power our cities could achieve with the return of great efficiency. In the long run, every city gains 10 to 30 percent per day from reducing leaks. This tremendous gain from capturing lost water first is redirected to municipalities' bottom line in new revenue, then everyone's efficiencies, and *eventually* other industries to lower and offset existing subsidy money sources; what was once lost can now be saved. Decreasing NRW directly lowers water inflation nationwide, thereby increasing the country's long-term efficiency, productive power, and buying power—putting water back into our money.

5. **The bond market becomes more opportunistic for the municipalities.** Higher water rates lower the municipalities' burden of bond sales to finance debt for CIPs. Naturally, the need for municipal bonds to fund infrastructure projects falls after water rates rise and water value is recognized. This decreases the weight

of the rating agencies such as Standard and Poor's. It empowers the municipalities' capacities to efficiently and internally finance local water and wastewater projects despite their city's previous performance. The demand for bonds initially falls, which decreases the cost of financing operations by lowering the cities' exposure to new debt. In essence, the effects decrease the net value of interest payments shelled out to bondholders. For decades, those bondholders would not get involved in failing cities with failing industries, unless higher interest rates incentivized them. But now, water and wastewater districts receive quality bondholders from the fundamental shift in higher water rates and in recognizing US water value. The demand from quality investors rises, reducing the interest required to entice *previous* would-be investors. The water rates increasing may cause our bill to rise in the short term; however, residents' exposure to principal and interest obligations of bonds on the water bill continuously fall. This can lower all water bills over time.

6. **There's no crisis, just happy corporations and citizens.** Because of higher water values, a natural increase in the public's expectation of rising prices in aqua-exports follow. Crises only occur without awareness. Otherwise, all risks can be hedged, costs planned around, and effects reduced. This allows for forecasting and prepares downstream producers and households for budgetary requirements in operations and at home. Businesses and residents can plan for higher prices. It's only unfortunate if we can't see inflation coming or don't know it's here— as we've experienced since the 1970s. Commodity prices will increase across agriculture and industry.

When aqua-exporters, like farmers, increase the value of their harvests, droughts become less devastating to the business model because farmers are making more on what they do yield. These prices are slowly implemented at the wholesale and retail level. Initially, inflation steadily rises through transparent price increases. This is a means to tackle deflationary cycles and focus on price stability (Federal Reserve tenets). Higher prices mean higher revenues for corporations, which spur investment in new jobs and education for existing employees, new manufacturing plants and equipment, and research and development positions. This money is reinvested into the US economy, causing macro-expansion in the water sector—exponential development. Higher corporate revenues and a rise in *stated* inflation increase the expectations for wage increases, and salary compensation becomes self-fulfilling, giving us a raise and leading to a series of raises.

7. **Economic balance will come from higher water rates.** If water prices rise through public water and wastewater operators, then consumer spending cannot significantly decrease across all market sectors. If some level of consumer spending declines, the offset in higher prices neutralizes the impact. Thus, no long-term recession or depression.

8. **Weather can no longer make it rain more or less money.** With higher water value, the effects of demand destruction (price collapse) and extreme price spikes *(overshooting effect)* can be controlled and minimized—reducing the volatility and noise around food and energy prices (Federal Reserve objectives for monetary policy). Higher prices in commodities

will not collapse again, as they have before, because prices of aqua-exports have a price floor in them, due to a supportive water value. *Prices will instead graduate.* Higher water rates create new price floors more consistently and responsibly than lobbying for our country's intervention and money to support profitable industries, a.k.a. subsidies or tax breaks. Higher water value offsets the secular impact of government spending to address changes ushered in by the Federal Water Pollution Control Act of 1948 and EPA Acts to the present. Additionally, this creates a pseudo-price ceiling, reducing the results of short-term volatility from inclement weather cycles, compensating for floods and droughts, containing the overshooting effect on price and businesses affected by water's force. For example, if flooding punishes US states responsible for the majority of corn, soybean, and wheat harvests, then the price of grains can skyrocket—but this could be prevented if we priced water into crops ahead of time. Additionally, in the long term, the investment in sewer system upgrades will protect the neighborhoods and businesses from floods and damages brought on by extreme heat, giving the housing market a raise.

9. **When you give, you get.** Higher water and wastewater rates mean more revenues from all businesses. This means more tax money paid to federal, state, and local governments without ever increasing tax rates. The value of tax receipts increases without anyone speaking and voting for taxes. Tax revenue grows and we benefit as PEs spill over.

10. **Who says you need a political party to reduce country debt?** Higher tax receipts reduce the deficits of federal, state, and local budgets. Surplus can return to our government's financial house.

11. **You get two raises. One now and one later.** Nominal wages increase due to consistent higher prices and steadily rising inflation. Wage increase is met and matched by the new employment demand from the national infrastructure effort funded by municipalities, which offsets the initial shock of higher prices as more citizens sitting on the sidelines enter the workforce.

12. **Transparency closes price mysteries.** The water gap shrinks as the price of goods and services rises to a true price structure. This reduces the need for intervention and subsidies which uphold industries protected by government. Higher-priced goods are better for our buying power in the long run, because we will benefit from the reduction in subsidy programs—an effort to hide "real" costs—and the value of our currency will strengthen from fundamental improvements in the country's financials (Brand USA).

13. **Make our water systems national treasures.** We have to protect water and wastewater districts. The pipes in our buildings made US history rich with possibilities. Maintaining and upgrading all pipes is a credit to our heritage and future, which require more accountability and attention than the buildings which stand on top of them. The cost of upgrading pipes is similar to the investment in preserving lands or historic sites. A higher water value, in conjunction with national treasure advocacy, enhances our families' water quality, enhances

their outputs (so they remain productive longer), and reduces negative externalities and ULs entering our waterways for bathing, drinking, farming, fishing, and swimming. In historic cities such as Annapolis, Norfolk, or Savannah, investments in sewer systems through increased rates for water and wastewater districts preserve the history and addresses the housing market. National treasure advocacy will stave off rising waters and reroute higher tides. The most effective way to address FEMA and private insurers, and strengthen the housing market is through the theory of the second best. Using this theory at the highest control center in the economic food chain—water—addresses the cure rather than the symptoms. National treasure status requires the sewer systems to be funded and enhanced in places like Baltimore, Maryland, Norfolk, Virginia, and Wilmington, Delaware, and upgraded in response to rising sea levels. Recognizing water value and funding the highest upstream option for *AIR* increases property values in cities and revitalizes crime-stricken communities with opportunities.

14. **Moratorium on privatization.** Place a moratorium on new private leases for operations and maintenance of local municipalities. Privatization is not the answer; it's an option with too many unknowns. Our water systems should be *Treated by America*.

15. **Don't tell me I can't have the filet and to have chicken instead.** Add a water quality index into the CPI and PCE. Base it on random aquifer, groundwater, lake, river, and well readings throughout each state and near industries. This addresses many undocumented concerns about the data used for the cost of living indexes and

inflation, thus securing all classes' progression at all ages. This give us better inflationary benchmarks. A decrease in the national water quality corresponds with an increase in the water quality weighted component of the CPI or PCE, increasing inflation and COLAs (cost of living adjustments). Conversely, when water quality increases (which will happen), that pushes down the CPI or PCE and benefits the US government and its payout of liabilities such as Social Security. This essentially puts water back in our money and incentivizes our government the right way. Water value then moves beyond its pure link to the price of commodities and that volatility. People can keep up with prices of everyday goods and services when water is in their wallets. They have the financial freedom to treat themselves to a steak, rather than chicken, or they can become vegans.

16. **We love choices. We can do so much with all that money.** Increasing water rates and channeling their return to replace tired infrastructure reverse the 10 to 30 percent losses from non-revenue water and become our gains. Such 10 to 30 percent gains responsibly aid industries with tangible social benefit rather than private sector industries with historical dependence. This redirects and revolutionizes subsidy responsibility from the US federal government and instead empowers the water industry to be a force in our fiscal direction (Norfolk's example of *local fiscal federalism*). Using our water supply as an economic force, we can take efficiency gains from revenues and reduce subsidies, lessen new tax policy considerations, and as we'll learn later, advocate potential trade policy.

Meanwhile, we can also pay our firefighters, police, teachers, and veterans.

17. **We are like little Walmarts with all that buying power.** The US dollar then grows stronger globally against other currencies. As you know, the benefits of higher water value shrink US deficits, and higher prices domestically signal very positive country fundamentals. Imports grow more affordable and offset export weakness. *International investors seek the US dollar for water value,* buy the currency, and invest in US and world stock markets using the dollar rather than their own currency. This is called the *carry trade,* where a trader exchanges a weaker currency to buy a stronger currency and uses that money to buy stocks in a country where its markets are promising. Therefore, a stock's value increases and the currency grows stronger. When the investor sells, their gains are realized from the stock and the US dollar's appreciation. Our well-deserved purchasing power grows.

18. **Another raise. The only thing that matters feels real.** Buying power grows strong for US citizens. Water inflation declines and ULs are plugged and on the mend. Thus, there is a *real* wage increase and savings realized. Productivity increases via water efficiency and water quality improvements for humans. Healthcare needs for both short- and long-term diseases diminish in the long run. This is a real raise.

19. **Savers finally get rewarded again.** Interest rates rise in response to higher commodity pricing and international investor demand turns bullish toward the fundamentals of the US economy and the dollar.

Because inflation causes the Fed's monetary policy to respond with higher interest rates, the dollar gains more momentum as investors bought the currency for yield (carry trade). After wages rise, so do GDP and gross national income (GNI). As the US economy heats up and GDP expands, interest rates rise again to cool down the economy, and the dollar rises again. In sum, interest rates graduate to an improved norm, rewarding savers.

20. **Investors get rewarded.** As GDP increases, so do the earnings of multinational corporations with US operations, and thus equities or stocks outperform. Stock markets rise in tandem with interest rates, because of higher recognized water value. We can invest in stocks or collect interest on savings. Equity markets don't have to be risky with interest rates rising, because the investment in higher water rates supports stock values. The equity markets rise because corporations are growing, economies are stable, inflation is steadily rising, employment is rising, and efficiencies are at optimum levels. Most dazzling of all, *the dollar, commodities, housing markets, interest rates, and US stock markets rise in unison in the long run.* Investors and 401(k) participants get paid.

21. **We can stabilize prices.** Demand destruction of aqua-exports does not occur with higher prices internationally, because commodity inflation is real and sustainable in the US—not climate-contingent and swinging wildly. There is short-term export pain, but the rest of the world does not back down from higher prices. Higher prices aren't a result of volatility in water supply levels due to drought or shortage (2008 is our example). After water rates in the US graduate, there

is a floor of support from water value. Commodities do not wildly fluctuate below that value or above that value. However, global inflationary trends need to be addressed soon thereafter. The US should not export inflation abroad when these fiscal changes occur.

22. **Exporting water value is a political token.** A new water value from the world leader serves as an equalizer for both deflation and inflation in developed economies. Water valued at proper price structures brings about economic equilibrium and market stability. The globe has experienced inflationary oil price spikes and deflationary collapses because water's force had been volatile, unsustainable, unrecognized by all, and unsupportive toward stable prices. History's peaks and troughs with oil prices occurred when a transparent water value had *not* been established in the public domain. Because the US dollar is invasive in global markets, a transparent US water value eventually brings stability in pricing aqua-exports globally. Exporting water value is a part of the international component and the path of leadership.

23. **Just like in the 1950s and 1960s, yes, we can be proud.** The US can once again be a global example: national economies can improve with higher valued water. The US leads developed nations and overcomes the recent tradition of Europe leading the environmental way. The US exports water value message to the globe and triggers an actionable platform to adapt to climate change. This gives the US more political tokens globally for international resolutions. Instead of play currency wars, the US instead reveals *current-cy* politics. While developing nations have larger gaps to fill in their

water value history, this is where the international policy subsidizes the change for lagging environmental countries. It provides a payoff for both developing nations and US citizens.

24. How to give the middle class a raise. There is now a leap in innovation, water savings, and technology. The service sector's response to higher prices from higher water value is met with solutions for less water-intensive processes and techniques for water-less production. Government desalination might make sense at these prices—when marine biologists and environmentalists can compromise. All innovation drastically reduces the overhead of the economy's water requirements. Increasing water efficiencies, reducing exposure to water expenses, and reducing leaks and using less water per good and service slowly lower overall costs and manage water's force on the market. This yields higher corporate revenues and profits and increases wages in tandem. Essentially, this is *how to give the middle class a raise.* Through engineering aqua-export prices with innovation, this naturally evokes educational environments. New job markets related to water efficiency departments and water services emerge. Service and technology expand and create efficient and therefore more profitable business models. Higher participation in education for water-less economies drives a new economy and new workforce. Corporate earnings, corporate profits, and investment returns grow and allow the US to become the digital hub for water valuation. Due to increased education and new water applications, corporate earnings outperform and stocks appreciate again and again from the total long-term investment. Life is promising, offering vast education opportunities, and criminal activity plaguing cities diminishes. Entitlement programs dissipate and dependency is phased out for lower socioeconomic classes.

25. **Decrease in capital improvement plans over the medium term.** Infrastructure dramatically improves in the short term and water quality rises. Slowly, a lessening infrastructure demand catches up with all capital improvement plans (CIPs) and EPA goals. As the CIP investment demand falls, the water value does not change, but the water rates plateau and serve other industries—for example, replacing subsidies at home for US operations abroad (*fiscal federalism through water, not through debt*)—or the water rate could fall. The water value of the country functions as a *current-cy* in terms of economic leadership and welfare. The process of recognizing water value *creates more returns for US interests and world*, rather than water's ownership and commodification *taking away from people*. Cost-benefit analysis confirmed.

26. **True cost structure.** Prices of goods and services reach their true cost structure and subsidies are relinquished and replaced for water revenues. We end up with a budget surplus.

27. **The water sector can subsidize.** Higher prices and revenues, efficiency gains, higher wages, more optimal employment, increased tax receipts without ever increasing tax rates, new budget surplus, education, and innovation further reduce the need for existing and new subsidies. New monies from water value and water revenue substantiate outside-the-box thinking for corporations.

CHAPTER 17

Quality Water for Every Tribe

THE AMERICAN PEOPLE FORMED a great nation through the use of global resources like water. And we now understand that global water inequality and gender inequality are mirror images of one another. Therefore, it benefits us to use water as a tool for international economic liberation—because it has always benefited us, and will always benefit us. When given an identity, water's force can change the way we look at the global environment for business.

Reforming how we value water through our exports and imports is a long-term goal that not only benefits the US, but could stabilize and uplift the rest of the world as well. The US has an opportunity to be a world leader in this regard, an opportunity we should seize. We could lead by example and show the benefits of recognizing water value: by charging higher water rates through government, upgrading our infrastructure, protecting our pipes, and revising inflation benchmarks like the CPI and PCE. We could show other countries how to find positive externalities in what appears to be negative externalities and unrecognized leaks (ULs). We could demonstrate tremendous gains in the US standard of living from increasing water rates for agriculture, industry, and residents and project our strength globally. We can partner with US corporations taking older models of subsidies outside the United States. We can offset domestic welfare and export water revenues to our corporate partners instead.

Once the world understands the United States has employed a water value policy, the US can lead other nations to roads never traveled before. It's time to cash in our political tokens for *water diplomacy*. Increasing water rates in the US through local municipalities is also why water rates or prices in underdeveloped countries could fall. If you could make that happen, why would you want to? Because benefiting your family and your country could also mean benefiting other economies. It could spawn trade agreements whereby we could receive drastic discounts on our imports from those economies as developing countries' water rates finally match their income levels. If we could benefit our families and those around the world, isn't that something we would want? Here's how.

As we increased the price of water rates over time in the US, we would realize that even the US dollar becomes an aqua-export, containing the valuable fundamentals that the US economy would now be exuding. Water value would fill the US dollar and present an opportunity to spin that buying power into a win-win-win. Yes, as US currency becomes stronger it lowers the world's demand for US exports. How could US citizens once again benefit and economically expand from this condition?

Just like having to pay more taxes is a good problem because it means you are making more money, the strength of the US dollar and higher export prices could be demonstrated to the world as a *good* problem, because it offers the American people options. And we love options. So how do we offset this dollar strength which would cause US exports to shrink? As we increase water value in the US through higher water rates, we can use the revenues from this fiscal policy and reinvest our surplus elsewhere, building strong economic ties to other aqua-exporting countries through our corporate partners.

Multinational corporations (MNCs) would praise this diplomacy. The US could make water investments internationally and justify specific policies because of US domestic water value gains.

Let's assume the new water value, via water rates, creates unimagined revenues and substantial new tax receipts, shrinks deficits, and eventually reduces entitlements and subsidies for US citizens and corporations. Water inflation would be lowered, ULs reduced through water quality improvement, and tremendous gains realized from the savings of non-revenue water. Once the US validates its water policy success, couldn't the government's fiscal policy focus on new channels for reinvestment aside from unnecessary domestic subsidies? Perhaps the water revenues would be attractive for our corporate operations abroad, who are exporting from emerging markets?

The lesson we could teach is that tax receipts would rise without ever increasing taxes, deficits and subsidies would shrink, and surpluses return to the US—all contributing to a shrinking water gap, fortifying families and their wallets. With these gains, US corporations would not truly lose their subsidy dollars, which could instead go to operations abroad that benefit the US economy. Why? Because the evaluation of economic returns from globalization has benefited US corporations, but not significantly improved the global economy, especially developing markets and their environments. For the United States to maintain Brand USA as its environmental regulations heated up after 1970, the developing world experienced many new market failures related to freshwater and stifled economic progress there. Similar to the repatriation of money to Germany after World War II, there's a historic need to repatriate value to developing countries since the EPA began. For there to be global economic stability, everyone must get access to basic needs so that they can contribute to a universal cause—a global economic growth model. The theory of the second best would allow for *extranational investment*. We would offshore water revenues to US-born corporations with operations abroad, generating a mutual benefit by subsidizing their infrastructure costs in underdeveloped markets. In return, US citizens and related corporations would receive a discount on imports from those specific relationships.

Poorer world economies have less water infrastructure, less access to water, and lower income-to-water rates than the US—when they live right next to freshwater. Developing nations pay ten times more for water than developed nations. For example, in "South Sudan, home to the Nile," an average family spends "a third of their earnings on water ... Many families, especially government employees, haven't received their salaries lately" and live on less than eight dollars a month with hyperinflation conditions, according to Reuters. Regardless of which nations we embrace, rebalancing water disparity can benefit the US. When we reach out and extend our hand, we call it *water diplomacy.*

To repeat: "Theory of the second best states that a hands-off policy is desirable in any one market only if all other markets are working properly. If they are not, a government intervention that appears to distort incentives in one market may actually increase welfare by offsetting the consequences of market failures elsewhere."

This theory justifies intervention based on the global market failure of freshwater brought on by globalization. US-based corporations could support the theory of the second best in two different continents via US corporate operations in developing countries. Cause and effect: When water rates rise in the US, water revenues and the price of everything taxed rises. With increasing tax receipts from higher water rates and higher corporate profits, subsidies can be decreased, and the US can truly recapture lost subsidy money, and thus reduce debt. The value of the dollar would rise. To offset this dollar appreciation, an increase in water diplomacy to developing countries and a discount sent home to the US would, in time, neutralize the US dollar's impact on American citizens and corporations. And through this exchange, a subsidy abroad would be able to lower the local cost of water for developing nations' citizens and increase their income-to-water rates. These global citizens would then be able to expand their buying power and capacity to live as domestic water rates fall in their country.

◇>$

Selected US-based multinationals with strategic business units in underdeveloped nations could set up manufacturing plants and equipment, and simultaneously invest in poverty-stricken rural and urban water infrastructure systems, which would benefit the people and the corporation's operations—both efficiency for higher profit margins and local environments. Water diplomacy would subsidize the infrastructure necessary for the operations of the manufacturing plants and the community's environmental efficiencies tied to that production.

Because of the positive externalities a US water value system reaps, corporations would take previous US monies directed at protecting industry at home, and transfer it to their foreign subsidiary, where they would use the dollars to invest in a community's infrastructure. This would create affordable water bills for local citizens linked to that industry. MNCs would be required to provide water and wastewater rates for a minimum number of citizens at costs that match their countries' wages—a cost reasonable to their income-to-water rate. A US water diplomat—hypothetically at the EPA—would review these plans.

Local governments in developing nations would hire employees, partner with EPA diplomats, and create an agency to monitor billing and infrastructure within those nations. The US would send *these* water diplomats to that country to oversee operations and maintenance in conjunction with local governments. Developing countries would get access to clean water, sanitation, and improved environments. This would initiate new investments in rural areas, creating the benefits of urban development such as hospitals and schools. Women and children could embrace an education rather than short-term walks for their families' water supplies or eliminate plaguing water-borne illnesses. Through education, gender equality could expand internationally, closing the chapter on the ripple effects of limited water resources on a country's societal norms. The benefit to the US is that imports from these production sites are purchased at a reduced margin to reward the US and its citizens for their investment in developing countries.

As subsidy trickles from US multinational corporations to their satellite sites in emerging markets, these strategic business units could absorb the subsidy benefits in underdeveloped nations where these MNCs would be operating. As part of the exchange, the new water infrastructure would create a water rate in line with the income of those citizens. Thus, MNCs would be unaffected by this exchange, because what is missing money in the US (a decrease in subsidy) is a gain abroad in money for plant and equipment (dollars created from the increase in water revenues in the US). These specific operations would be responsible for the discounted goods sold to the US justifying the investment, creating enhanced profit margins for corporations, and water diplomacy for those developing countries.

Finding water value and closing the water gap is *not* a zero-sum game. We all benefit and, by so doing, give ourselves a real raise in buying power and a better shot at attaining the American dream—a global dream.

CITATIONS AND SOURCES

My conclusions and views in *Waconomics* are based upon extensive research since 2003. Below is a list of the main sources of information quoted, referred to, or cited in this work.

SOURCES

"3 Major Water Main Breaks Reported In Bayonne, NJ." June 13, 2017. CBS New York. Accessed October 20. http://newyork.cbslocal. com/2017/06/13/bayonne-water-main-breaks/.

Adams, Francis. Class Notes, Political Science 337. Spring 2006 and 2011. Old Dominion University.

Al-Attiya. "A country with no water." TED Talks. https://www. ted.com/talks/fahad_al_attiya_a_country_with_no_water? language=en#t-470739.

Allen, Gary. *The Rockefeller File*. Seal Beach, CA: '76 Press, 1976.

Amadeo, Kimberly. "Balance of Trade: Favorable vs Unfavorable." About. com. June 1, 2017. http://useconomy.about.com/od/tradepolicy/g/ Balance-of-Trade.htm.

American Association for Justice. "The Ten Worst Insurance Companies in America." American Association for Justice. https://www. justice.org/sites/default/files/file-uploads/AAJ_Report_ TenWorstInsuranceCompanies_FINAL.pdf.

Amici, Randy. Personal interview. July 9, 2008.

"Annapolis." National Trust for Historic Preservation. http://savingplaces. org/treasures/annapolis#.VVoJX2BN38s.

"Arsenic in Your Juice." *Consumer Reports.* January 2012. http://www. consumerreports.org/cro/magazine/2012/01/arsenic-in-your-juice/ index.htm.

Ari, Ismu Rini Dwi. "Non-Revenue Water." International Water Association. http://www.iwahq.org/1ny/themes/managing- utilities/utility-efficiency/non-revenuewater.html.

Associated Press. "Foreign companies buying U.S. roads, bridges." *USA Today.* July 15, 2006. http://usatoday30.usatoday.com/news/ nation/2006-07-15-u.s.-highways_x.htm.

"Awful Weather We're Having." *The Economist.* September 30, 2004. http://www.economist.com/node/3254119.

Bacon, Charlotte. *Split Estate.* New York, NY: Picador, 2007.

Bancalari, Kellie. "Private college tuition is rising faster than inflation… again." *USA Today.* June 9, 2017. http://college.usatoday. com/2017/06/09/private-college-tuition-is-rising-faster-than- inflation-again/

Balaam, David N. and Michael Veseth. *Introduction to International Political Economy*. 2nd ed. Upper Saddle River, NJ: Prentice-Hall, 2001.

Barlow, Maude and Sam Bozzo (dir.). *Blue Gold: World Water Wars*. Purple Turtle Films, 2009.

Bartlett, Christopher, Sumantra Ghoshal, and Paul Beamish. *Transnational Management: Text, Cases, and Readings in Cross Border Management*. 5th ed. New York, NY: McGraw-Hill/Irwin, 2008.

Baye, Michael R. *Managerial Economics and Business Strategy*. 6th ed. New York, NY: McGraw-Hill/Irwin, 2009.

"Bechtel." Wikipedia. Accessed March 13, 2010. https://en.wikipedia.org/wiki/Bechtel.

Blouet, Brian W. and Olwyn M. Blouet. *Latin America and the Caribbean: A Systematic and Regional Survey*. 6th ed. New York, NY: John Wiley, 2010.

Bottemiller, Helena. "Study Finds Arsenic in Apple Juice." http://www.foodsafetynews.com/2010/03/study-finds-concerning-levels-of-arsenic-inapple-juice/#.VTUzA85N38s.

Brock, James. *The Structure of American Industry*. 12th ed. Upper Saddle River, NJ: Pearson Education, Inc., 2009. 155–182.

"Brock News." Brock Associates Website. Accessed October 26, 2012. https://brocku.ca/brock-news/.

The Brock Report. Brock Associates, April 19, 2013.

The Brock Report. Brock Associates, August 10, 2012.

The Brock Report. Brock Associates, June 29, 2012.

Brodrick, Sean. "A Dangerous Mix of Water and Oil." *Uncommon Wisdom.* Accessed March 19, 2010. https://www.uncommonwisdomdaily. com/a-dangerous-mix-of-water-and-oil-6-9001.

Brown, Lester R. *Plan B 3.0: Mobilizing to Save Civilization.* New York, NY: W.W. Norton & Company, 2008.

Bruce, Jim and Live Schreiber (dir.). *Money for Nothing: Inside the Federal Reserve.* Virgil Films and Entertainment, 2013.

Buncher, Ryan. "United Water Rate Hike Begins." *Patch.* August 26, 2015. http://pearlriver.patch.com/groups/politics-and-elections/p/ municipalities-challenge-ofunited-water-rate-hike-begins.

Bureau of Labor Statistics. "Consumer Price Index Frequently Asked Questions." Bureau of Labor Statistics. http://www.bls.gov/cpi/ cpifaq.htm.

Bureau of Labor Statistics. "CPI Addendum to Frequently Asked Questions." Bureau of Labor Statistics. http://stats.bls.gov/cpi/ cpiadd.htm#4_1.

CALIFORNIA AGRICULTURE: Feeding the Future 2003. Sacramento, CA: Governor's Office of Planning and Research, Vital Communities Institute, California Rural Community Task Force, 2003.

"California Slips to Number 9 in World Economic Rankings." *Sacramento Bee.* January 2012. http://blogs.sacbee.com/ capitolalertlatest/2012/01/california-slips-to-number-9-in- worldeconomic-rankings.html.

"Cambodia." *Wikipedia.* Accessed April 7, 2010. https://www.wikipedia. org/Cambodia.

Campus Editor. "California Agriculture Production Ranks in Top 10 Worldwide." *The Aggie.* April 3, 2008. http://www.theaggie. org/2008/04/03/california-agriculture-productionranks-in-top-10worldwide/.

Carroll, Joe. "Worst Drought in More Than a Century Strikes Texas Oil Boom." *Bloomberg News.* Accessed June 13, 2011. https://www. bloomberg.com/news/articles/2011-06-13/worst-drought-in-more-than-a-century-threatens-texas-oil-natural-gas-boom.

Caruso-Cabrera, Michelle. "Chile's Water Real Estate." CNBC. September 30, 2010. http://video.cnbc.com/gallery/?video=1603801303.

Casselman, Ben and Phil Izzo. "Number of the Week: U.S. Producing More Oil Than It Imports." November 16, 2013. http://blogs.wsj. com/economics/2013/11/16/number-of-the-week-u-s-producing-more-oilthan-it-imports/.

"Chile, Water Privatization Showcase." Food and Water Watch. http:// www.foodandwaterwatch.org/global/latin-america/chile/chile-%E2%80%93-waterprivatization-showcase/.

Chossudovsky, Michel. "Chile, September 11, 1973: The Ingredients of a Military Coup." *GlobalResearch.* Accessed March 2, 2010. https://www.globalresearch.ca/chile-september-11-1973-the-ingredients-of-a-military-coup-the-imposition-of-a-neoliberal-agenda/5545251.

Cleaves, Peter S. *Politics and Administration in Chile.* Berkeley and Los Angeles, CA: University of California Press, 1974.

"Climate Change Water Infrastructure." *Water Finance and Management Journal.* http://uimonline.com/index/webapp-stories-action/ id.1317/title.impact-climate-changewater-infrastructure.

Collins, Joseph and John Lear. *Chile's Free-Market Miracle: A Second Look.* Oakland, CA: Food First, 1995.

"CPI Defined." Investopedia. http://www.investopedia.com/terms/c/ consumerpriceindex.asp.

"Currency." Investopedia. http://www.investopedia.com/terms/c/currency. asp.

Curzio, Frank. "The Jim Rogers Interview You Don't Want to Miss." Stansberry and Associates Radio Broadcast. Podcast audio. July 20, 2011. http://www.thedailycrux.com/Post/37891/the-jim-rogers-interview-you-don-t-want-tomiss.

David, Cristina C. "MWSS Privatization: Implications on the Price of Water, the Poor, and the Environment." *Philippine Institute for Development Studies.* Discussion Paper Series No. 2000–14. 2000. 7.

"Demand Destruction." Business Dictionary. http://www. businessdictionary.com/definition/Demand-destruction.html.

Diamond, Jared and Cassian Harrison (dir.). *Guns, Germs, and Steel.* National Geographic, 2005.

Dietz, David and Darrell Preston. "The Insurance Hoax." *Bloomberg News.* Accessed September 2007. https://www.bloomberg.com/ news/marketsmag/mm_0907_story1.html.

"Disaster and Its Shadow." *The Economist,* September 14, 2002.

"Disease." Water.org. http://water.org/water-crisis/water-facts/disease/.

Drew, Gina and Mike Levien. "The Opposition to Coca Cola and Water Privatization: Activists in Medhiganj, India Rise Up." *Z Communications*. Accessed July 25, 2006. http://www.zmag.org/content/showarticle.cfm?SectionID=66&ItemID=10641.

"Drinking Water Requirements for States and Public Water Systems." Environmental Production Agency. http://water.epa.gov/lawsregs/rulesregs/sdwa/gwr/basicinformation.cfm.

Easterly, William. *The Elusive Quest for Growth: Economists' Adventures and Misadventures in the Tropics.* Cambridge: MIT Press, 2001.

"Economics." Water.org. http://water.org/water-crisis/water-facts/economics/.

"Economies of Scale." Investopedia. http://www.investopedia.com/terms/e/economiesofscale.asp.

"Economy/Prices." 1970s Flashback. http://www.1970sflashback.com/1977/economy.asp.

Ellis, Mark, Sarah Dillich, and Nancy Margolis. "Industrial Water Use and Its Energy Implications." Office of Energy Efficiency and Renewable Energy. http://www1.eere.energy.gov/manufacturing/resources/steel/pdfs/water_use_rpt.pdf.

"Engineering Solutions." Water for All. http://12.000.scripts.mit.edu/mission2017/solutions/engineering-solutions/.

"EPA—Natural Gas." Environmental Protection Agency. http://www.epa.gov/cleanenergy/energy-and-you/affect/natural-gas.html.

Espenshade, T. J., J. C. Guzman, and C. F. Westoff. "The surprising global variation in replacement fertility." *Population Research and Policy Review* 22, no. 5/6 (2003): 575. doi:10.1023/B:POPU.0000020882.29684.8e.

"Fact #915: March 7, 2016 Average Historical Annual Gasoline Pump Price, 1929–2015." Office of Energy Efficiency and Renewable Energy. http://energy.gov/eere/vehicles/fact-835-august-25-average-historical-annual-gasolinepump-price-1929-2013.

Ffrench-Davis, Ricardo. "Export Dynamism and Growth in Chile Since the 1980s." *CEPAL Review* 76 (2002): 1–19.

"Fiat Money." Investopedia. http://www.investopedia.com/terms/f/fiatmoney.asp.

Fisher, James. "A Sandy headache that won't go away." *Delawareonline.* November 06, 2015. http://www.delawareonline.com/story/news/local/2015/11/06/sandy-headache-wont-go-away/75058700/.

Flanagan, Brenda. "Bayonne Water Rates Spike After Privatization." January 5, 2017. NJTV News. Accessed March 4, 2018. https://www.njtvonline.org/news/video/bayonne-water-rates-spike-privatization.

Fluence News Team. "What is non-revenue water?" *Fluence.* February 29, 2016. https://www.fluencecorp.com/what-is-non-revenue-water/.

Fontevecchia, Agustino. "Bernanke Fights Ron Paul In Congress: Gold Isn't Money." http://www.forbes.com/sites/afontevecchia/2011/07/13/bernanke-fights-ron-paul-incongress-golds-not-money/.

Food and Water Watch. "Water Privatization: Facts and Figures." *Food and Water Watch*. Accessed November 19, 2008. https://www. foodandwaterwatch.org/insight/water-privatization-facts-and-figures.

Food and Water Watch. United Water: SUEZ Environnement's Poor Record in the United States. Accessed May 2010. http://www. foodandwaterwatch.org/reports/united-water/.

Fordham University "Hogan Steel Archive." *Fordham University Libraries.* Accessed April 1, 2010. http://www.library.fordham.edu/archives/ hogan.html.

"Forex - FX." Investopedia. http://www.investopedia.com/terms/f/forex. asp.

"Forex Fundamentals." Investopedia. http://www.investopedia.com/ terms/l/liquidmarket.asp.

Fox, Josh, Frederic P. Miller, Agnes F. Vandome, and John McBrewster. *Gasland.* Alphascript Pub., 2010.

Foxwood, Naomi. "Making Every Drop Count." *Water Finance and Management Journal.* http://uimonline.com/index/webapp-stories-action/id.922/title.making-every-drop-count.

"Free Market." Investopedia. http://www.investopedia.com/terms/f/ freemarket.asp.

"Free market." Wikipedia. https://en.wikipedia.org/wiki/Free_market.

Frum, David (2000). *How We Got Here: The '70s.* New York, NY: Basic Books.

Fulton, Julian. "With Water, California's Bigfoot Is Imported." Pacific Institute Insights. March 10, 2014. http://pacinst.org/with-water-californias-bigfoot-is-imported/.

"GDP." Investopedia. http://www.investopedia.com/university/releases/gdp.asp.

Glanz, James. "Report on Iraq Security Lists 310 Contractors." *The New York Times.* October 29, 2008. http://www.nytimes.com/2008/10/29/world/middleeast/29protect.html?_r=0.

Glinski, Stefanie. "The price of water: South Sudan's Capital Goes Thirsty as Costs Soar." Reuters. September 30, 2017. http://www.reuters.com/article/us-southsudan-water/the-price-of-water-south-sudans-capital-goes-thirsty-as-costs-soar-idUSKCN1C508J.

"Global Risks 2014." World Economic Forum. http://www3.weforum.org/docs/WEF_GlobalRisks_Report_2014.pdf.

"Globalization." Dictionary.com. http://dictionary.reference.com/browse/globalization.

"Globalization." Investopedia. http://www.investopedia.com/terms/g/globalization.asp.

Goodfriend, Marvin. "The Phases of U.S. Monetary Policy, 1987 to 2001." http://www.richmondfed.org/publications/research/economic_quarterly/2002/fall/pdf/ goodfriend.pdf.

Gordon, Robert J. "The Boskin Commission Report: A Retrospective One Decade Later." NBER Working Paper 12311, June 2006. http://www.nber.org/papers/w12311.pdf?new_window=1.

"Gross Domestic Product." Investopedia. http://www.investopedia.com/terms/g/gdp.asp.

Gross, Terry. "Drought in California Creates War Between Farmers, Developers, Residents." NPR. April 30, 2015. http://www.npr.org/2015/04/30/403283276/drought-in-calif-creates-water-wars-betweenfarmers-developers-residents.

"Groundwater Quality in Principal Aquifers." National Water-Quality Assessment by US Department of Interior and US Geological Survey http://wawter.usgs.gov/nawqa/pubs/prin_aq/.

Hall, Sarah. Personal interview. December 3, 2013.

Hausman, Jerry and Ephraim Leibtag. "Consumer Benefits from Increased Competition in Shopping Outlets: Measuring the Effect of Wal-mart," NBER Working Paper 11809, December. http://www.nber.org/papers/w11809.pdf.

Hauter, Wenonah. "Ten Years Later, the 'Halliburton Loophole' and America's Dirty Fracking Boom." Food and Water Watch. August 10, 2015. https://www.foodandwaterwatch.org/news/ten-years-later-halliburton-loophole-and-americas-dirty-fracking-boom.

Heberger, Matthew and Kristina Donnelly. "OIL, FOOD, AND WATER: Challenges and Opportunities for Califronia Agriculture." Pacific Institute. December 2015. http://pacinst.org/wp-content/uploads/2015/12/PI_OilFoodAndWater_.pdf.

Helferich, Gerard. "Cry Me a Tributary." *The Wall Street Journal.* November 22, 2012. https://www.wsj.com/articles/SB10000872396390444897304578046752150779148.

"History of the Clean Water Act." Environmental Protection Agency. http://www2.epa.gov/laws-regulations/history-clean-water-act.

"History." SQM. http://www.sqm.com/en-us/acercadesqm/
informacioncorporativa/historia.aspx.

Hitchens, Christopher and Eugene Jarecki (dir.). *The Trials of Henry Kissinger*. First Run Features, 2002.

Hodge, Nick. "Blue Gold: Profitable Water Solutions." *Gold World*. Accessed April 15, 2009. http://financialmarket99.blogspot. com/2009/04/blue-gold-profitable-water-solutions.html.

Hodge, Nick. "Desalination Companies as an Energy Play: The Commodity All Energy Technologies Need." *Energy and Capital*. Accessed March 3, 2010. https://www.energyandcapital.com/ articles/desalination-companies-stock/1089.

Hodge, Nick. "Investing in Water: An Absolute Right to Profit." *Energy and Capital*. Accessed April 17, 2009. https://www.energyandcapital. com/articles/investing-water/863.

Hodge, Nick. "Profiting from Big Oil's Second Act." *Gold World*. Accessed April 17, 2009. http://financialmarket99.blogspot.com/2009/04/ profiting-from-big-oils-second-act.html.

"How does an expense affect the balance sheet?" Accounting Coach. http://www.accountingcoach.com/blog/expense-affect-balance- sheet.

"How much petroleum does the United States import and export?" EIA. http://www.eia.gov/tools/faqs/faq.cfm?id=727&t=6.

Hughes, J. David. "Drill, Baby, Drill." Shale Bubble. http://shalebubble. org/drill-baby-drill/.

"IMF says energy subsidized by \$5.3 trillion worldwide." CNBC. http:// www.cnbc.com/id/102688969.

"Imports." Investopedia. http://www.investopedia.com/terms/i/import.asp.

"Inflation Defined." Investopedia. http://www.investopedia.com/terms/i/inflation.asp.

"ITT Corporation." Wikipedia. Accessed March 13, 2010. https://en.wikipedia.org/wiki/ITT_Corporation.

Kaplan, Robert. "Was Democracy Just a Moment?" *The Coming Anarchy: Shattering the Dreams of the Post Cold War.* New York, NY: Random House, 2000. 59–98.

Kaplan, Robert. *The Ends of the Earth: A Journey to the Frontiers of Anarchy.* New York, NY: Vintage Books, 1996.

Kerr, Kevin. "The world's most important commodity is often the most overlooked!" *Money and Markets.* Accessed July 6, 2011. https://www.moneyandmarkets.com/the-world%E2%80%99s-most-important-commodity-is-often-the-most-overlooked-45676.

Koba, Mark. "Consumer Price Index: CNBC Explains." CNBC. Accessed August 4, 2011. http://www.cnbc.com/id/43769766.

Korte, Gregory and Ian James. "White House launches 'moonshot for water.'" *USA Today.* December 15, 2015. https://www.usatoday.com/story/news/politics/2015/12/15/obama-administration-launches-all-out-push-water/77356070/.

Krugman, Paul and Maurice Obstfeld. *International Economics: Theory and Policy.* 8th ed. Boston: Pearson Education, Inc., 2009.

Leithead, Alistair. "Exporting Alfalfa." BBC. February 19, 2014. http://www.bbc.com/news/magazine-26124989.

Long, Heather. "Where Harvey is hitting hardest, 80 percent lack flood insurance." *Washington Post.* August 29, 2017. https://www.washingtonpost.com/news/wonk/wp/2017/08/29/where-harvey-is-hitting-hardest-four-out-of-five-homeowners-lack-flood-insurance/?utm_term=.33c024fed743.

Lydon, Tom. "Investors Gravitating to Water ETFs." *ETF Trends.* Accessed May 18, 2008. http:// www.etftrends.com/2008/05/investors-dive.html.

Magyer, Joe. "The Greatest Company in the History of the World." *The Motley Fool.* Accessed April 8, 2009. https://www.fool.com/investing/dividends-income/2009/04/08/the-greatest-company-in-the-history-of-the-world.aspx.

"Martin Bormann." Britannica. https://www.britannica.com/biography/Martin-Bormann.

Mather, Mark. "World Population Data Sheet 2014: The Decline in U.S. Fertility." Population Resource Bureau. December 2014. http://www.prb.org/Publications/Datasheets/2014/2014-world-population-data-sheet/us-fertility-decline-factsheet.aspx.

Mattera, Philip. "Subsidizing the Corporate One Percent." Good Jobs First. February 2014. http://www.goodjobsfirst.org/sites/default/files/docs/pdf/subsidizingthecorporateonepercent.pdf.

McConnell, Campbell R. and Stanley L. Brue. *Economics: Principles, Problems, and Policies.* 17th ed. New York, NY: McGraw-Hill/Irwin, 2008.

McCoy, Terrence. "Can China clean up its pollution before it's too late?" *The Washington Post.* June 4, 2014. http://www.washingtonpost.com/news/morning-mix/wp/2014/06/04/can-china-clean-upits-pollution-before-its-too-late/.

McCracken, Michael W. "Should Food Be Excluded from Core CPI?" *National Economic Trends.* http://research.stlouisfed.org/publications/net/20110801/cover.pdf.

McDiarmid, Margo. "Natural Resource Trouble in Alberta Canada." The Progress Report. http://www.progress.org/water16.htm.

McDonald, Steve. "'Slap in the Face' Award: How the Fed Cost Investors $500 Billion." Wealthy Retirement. April 10, 2015. http://wealthyretirement.com/fed-five-hundred-billion-dollars-unearned-income/?src=email.

McDonald, Steve. "'Slap in the Face' Award: The Fed's Flat-Out Lying About Inflation." Wealthy Retirement. March 27, 2015. https://wealthyretirement.com/slap-in-the-face-award-video/inflation-federal-reserve-fed-lying/.

McDonald, Thomas. Personal interview. May 1, 2009.

McElwee, Sean. "These five charts prove that the economy does better under a Democratic president." *Salon.* December 28, 2015. http://www.salon.com/2015/12/28/these_5_charts_prove_that_the_economy_does_better_under_democratic_presidents/.

Millard, Candice. *The River of Doubt: Theodore Roosevelt's Darkest Journey.* New York, NY: Doubleday, 2005.

Miller, Michelle. "Detroit water shut-offs." CBS. October 20, 2014. http://www.cbsnews.com/news/detroit-water-shut-offs-brings-u-n-scrutiny/.

Mis, Magdalena. "The world will face a massive water shortage in 15 years unless we start changing our consumption habits." *Business Insider.* Reuters. March 20, 2015. http://www.businessinsider.com/r-business-as-usual-will-create-a-thirsty-planet-in-15-years-says-un-2015-3.

"Monetize." Investopedia. http://www.investopedia.com/terms/m/monetize.asp.

Montagne, Renee. "Chile: 129 to Be Arrested in 'Dirty War' Crimes." NPR. http://www.npr.org/templates/story/story.php?storyId=112460666.

Monthly Brock Crop Report. Brock Associates, November 2012.

Moody's Investor Service. Data on Munich Re, Hanover Re, and Swiss Re, at www.moodys.com. Accessed July 26, 2007. URLs available to subscribers only.

"Moody's Downgrades Munich Re's Ratings to 'Aa1,'" *Insurance Journal.* September 20, 2002.

Moors, Kent. "Three Hidden Water Costs That May Already Be Boosting Energy Prices." *Oil & Energy Investor.* Accessed May 28, 2013. https://www.streetwisereports.com/pub/na/three-hidden-water-costs-that-may-already-be-boosting-energy-prices.

Moran, Tara. "The Hidden Costs of Groundwater Overdraft." Stanford. http://waterinthewest.stanford.edu/groundwater/overdraft/ .

Mshana, Rogate R. "Water, the source of life: Preservation, responsible management and equitable distribution: the ecumenical perspective." *Oikoumene.* Accessed December 9, 2008. http://www.wcc-coe.org/wcc/what/jpc/water-preservation.html.